MEMOIRS
of the
American Mathematical Society

Number 842

The Calculus of One-Sided *M*-Ideals and Multipliers in Operator Spaces

David P. Blecher
Vrej Zarikian

January 2006 • Volume 179 • Number 842 (first of 5 numbers) • ISSN 0065-9266

American Mathematical Society
Providence, Rhode Island

2000 *Mathematics Subject Classification.*
Primary 46L07, 46L89; Secondary 46B20, 46B04.

Library of Congress Cataloging-in-Publication Data

Blecher, David P., 1962–
    The calculus of one-sided M-ideals and multipliers in operator spaces / David P. Blecher, Vrej Zarikian.
        p. cm. — (Memoirs of the American Mathematical Society, ISSN 0065-9266 ; no. 842)
    "Volume 179, number 842 (first of 5 numbers)."
    Includes bibliographical references.
    ISBN 0-8218-3823-7 (alk. paper)
    1. Operator algebras. 2. Operator spaces. 3. Operator ideals. I. Zarikian, Vrej, 1972– .
II. Title. III. Series.

QA3 .A57   no. 842
[QA326]
510 s—dc22   2005053579
[512′.556]

Memoirs of the American Mathematical Society

This journal is devoted entirely to research in pure and applied mathematics.

Subscription information. The 2006 subscription begins with volume 179 and consists of six mailings, each containing one or more numbers. Subscription prices for 2006 are US$624 list, US$499 institutional member. A late charge of 10% of the subscription price will be imposed on orders received from nonmembers after January 1 of the subscription year. Subscribers outside the United States and India must pay a postage surcharge of US$31; subscribers in India must pay a postage surcharge of US$43. Expedited delivery to destinations in North America US$35; elsewhere US$130. Each number may be ordered separately; *please specify number* when ordering an individual number. For prices and titles of recently released numbers, see the New Publications sections of the *Notices of the American Mathematical Society*.

Back number information. For back issues see the *AMS Catalog of Publications*.

Subscriptions and orders should be addressed to the American Mathematical Society, P. O. Box 845904, Boston, MA 02284-5904, USA. *All orders must be accompanied by payment.* Other correspondence should be addressed to 201 Charles Street, Providence, RI 02904-2294, USA.

Memoirs of the American Mathematical Society is published bimonthly (each volume consisting usually of more than one number) by the American Mathematical Society at 201 Charles Street, Providence, RI 02904-2294, USA. Periodicals postage paid at Providence, RI. Postmaster: Send address changes to Memoirs, American Mathematical Society, 201 Charles Street, Providence, RI 02904-2294, USA.

The Calculus of One-Sided *M*-Ideals and Multipliers in Operator Spaces

Contents

Abstract

The theory of one-sided M-ideals and multipliers of operator spaces is simultaneously a generalization of classical M-ideals, ideals in operator algebras, and aspects of the theory of Hilbert C^*-modules and their maps. Here we give a systematic exposition of this theory. The main part of this memoir consists of a 'calculus' for one-sided M-ideals and multipliers, i.e. a collection of the properties of one-sided M-ideals and multipliers with respect to the basic constructions met in functional analysis. This is intended to be a reference tool for 'noncommutative functional analysts' who may encounter a one-sided M-ideal or multiplier in their work.

Received by the editor Oct. 19, 2003.

1991 *Mathematics Subject Classification.* Primary 46L07; 46L89 Secondary 46B20, 46B04.

Key words and phrases. Operator spaces, completely bounded maps, M-ideals, ideals in operator algebras.

Blecher was supported by a grant from the NSF.

Zarikian was supported by an NSF VIGRE Postdoctoral Fellowship at the University of Texas.

CHAPTER 1

Introduction

One of the goals of 'noncommutative functional analysis' is to find effective 'noncommutative variants' of important classical ('commutative') tools and theories. In this memoir, we establish an ambitious program proposed by E. G. Effros in 2000, and are able to accomplish a more or less wholesale importation of the basic theory of M-ideals into the noncommutative realm of operator spaces.

The classical theory of M-ideals emerged in the early seventies with the paper [**AE**] of Alfsen and Effros, and is now a important tool in functional analysis (see [**HWW**] for a comprehensive treatment, and for references to the extensive literature on the subject). Recently, in work of the authors together with Effros [**BEZ**], and in the second author's Ph.D. thesis [**Z1**], a one-sided variant of this classical theory was developed. The intention was to create a tool for 'noncommutative functional analysis', where one might expect 'ideals' to come in two varieties: 'left' and 'right'. Our one-sided theory contains the classical M-ideals and summands as particular examples, as well as the important class of *complete M-ideals* introduced about a decade ago by Effros and Ruan [**ER1**]. Other interesting examples of one-sided M-ideals include right ideals in C^*-algebras, submodules of Hilbert C^*-modules, and an important class of right ideals in nonselfadjoint operator algebras (see also [**BSZ**]). In this memoir, we give a systematic and up-to-date account of the 'one-sided M-theory'; in particular we exhibit here the 'calculus of one-sided M-ideals'.

One major device used in the analysis of one-sided M-ideals, and which is quite useful in its own right, is the notion of one-sided operator space multipliers introduced in [**B6**]. We take the theory of these multipliers much further here. The 'left M-projections' are exactly the orthogonal projections in a certain algebra $\mathcal{A}_\ell(X)$ of such multipliers. In fact, $\mathcal{A}_\ell(X)$ is a C^*-algebra, and moreover is a von Neumann algebra if X is a dual operator space. This is a key point in our theory, since amongst other things, it allows us to manipulate left M-projections in the way that operator theorists and C^*-algebraists are used to. Applying basic von Neumann algebra theory leads to our most interesting results.

We emphasize that our results by no means give useful conclusions for every operator space. Indeed, just as is the case in the classical M-ideal theory, a poorly chosen, or completely randomly chosen, space will have no nontrivial one-sided M-ideals or multipliers. As in the classical case, our one-sided theory detects only specific—but important—structure which a space may or may not have. This structure may loosely be called 'residual column-sum (or row-sum) structure', or 'residual operator algebraic structure'. See e.g. [**BZ, B4**] for an amplification of this point. Spaces like the predual of a von Neumann algebra do not have such structure, nor should they be expected to. Nonetheless, many operator spaces of interest do have such structure; indeed this structure arises naturally in many diverse settings. Moreover, one-sided M-ideals and multipliers have already proved their worth in the literature. They have birthed, for example, much of the current research on general operator algebras, such as

- the final form of the abstract characterization of dual operator algebras [**BLM, BM**];
- Kaneda and Paulsen's quasimultiplier variant of one-sided multipliers [**KP**], which yields a characterization of nonunital operator algebras [**Kan**];
- the first theory of ideals in general nonselfadjoint operator algebras (see e.g. [**B5**] and Section 4.5 below, where it is shown that the results in the present paper are key to the study of a large class of such ideals).

As another example, the one-sided M-ideal and multiplier theory has yielded several deep general results in the theory of operator bimodules, where one-sided M-ideals and multipliers frequently arise quite naturally (see e.g. [**BLM**] and references therein). In particular, they are a powerful tool in the study of *dual* operator bimodules (see e.g. [**BEZ, B3**]). Indeed, the theory has proved particularly effective at resolving delicate questions involving duality, and promises to continue to do so (see e.g. [**BM**]).

Thus it is quite certain that, just as in the classical case, one-sided M-ideals and multipliers will continue to arise naturally, from time to time, in connection with certain questions involving operator algebras, operator spaces, and operator bimodules. The main function of this memoir is to serve as a repository for the basic 'calculus' of these objects. We have intentionally written a reference tool for a reader who might encounter a one-sided multiplier or M-ideal in their work, carefully collecting here the properties of one-sided M-ideals and multipliers, and how they behave with respect to various constructions. Many of our results consist of 'one-sided counterparts' to the classical calculus of M-ideals of Banach spaces, and to the related theory of multipliers and centralizers of such spaces (which the reader will find comprehensively treated in [**Beh1, HWW**]). The proofs, however, are far from being simple imitations of the classical ones, and require quite different, 'noncommutative', arguments. Other results which we obtain appear to be new even for classical M-ideals, or at any rate we were not able to find them in the literature. We exhibit many results which are truly 'noncommutative' in nature, having no classical counterpart (e.g. the existence of one-sided M-ideals in certain noncommutative L^p spaces). We give a large assortment of examples coming from rather diverse sources (e.g. Hilbertian operator spaces, Hilbert C^*-modules, nonselfadjoint operator algebras, and low finite-dimensional operator spaces). These examples show that the noncommutative versions of approximately half of the classical 'calculus of M-ideals' break down without further hypotheses. For example, the set of one-sided M-ideals in a given operator space is closed with respect to the 'closed span' operation, $\vee$, but is not closed with respect to intersection, $\wedge$. The reason for this is related to Akemann's result that the 'meet' of two open projections need not be open [**Ake1**]. The intersection of two one-sided M-ideals *is* a one-sided M-ideal if one imposes certain conditions considered by Akemann. In summary, we are able to successfully 'quantize' almost all of the basic classical 'calculus of M-ideals'.

We now describe, very briefly, the structure of this memoir. In Chapter 2, we give definitions and basic results that will be used throughout. In Chapter 3, we study how projections and partial isometries in the abstract multiplier algebras manipulate the underlying space X. The basic von Neumann algebra projection calculus is key here. Chapter 4 is devoted to examples. The lengthy Chapter 5 is the apex of our work, and is intended as a systematic presentation of the basic calculus of one-sided M-ideals and multipliers. We discuss subspaces, quotients, duality, interpolation, tensor products, etc. Chapter 6 treats type decomposition and Morita equivalence for dual operator spaces. The short Chapter 7 is devoted to the complete M-ideals of Effros and Ruan (see [**ER1**]), and the associated 'centralizers'. This is very similar to the classical theory, however we deduce the main results quickly from results in earlier chapters. In Chapter 8 we discuss a few directions for future progress. In the Appendices we list some background facts which are used very often, about Banach spaces and infinite matrices over operator spaces.

We end this introduction with a little notation. Other notation will be encountered as we proceed. The underlying field is always $\mathbb{C}$. We usually use the letters $H, K, L, ...$ for Hilbert spaces. We write $\overline{S}$ for the *norm closure* of a set S, and $\overline{S}^{\mathrm{wk}^*}$ for the *weak-$*$ closure*. Notationally, a problem arises in the use of the symbol $*$, which we unfortunately use for three different things. Namely, we have the dual space X^* of a space X, the adjoint or involution S^* of an operator on a Hilbert space, and the adjoint operator $R^* : Y^* \to Z^*$ of an operator $R : Z \to Y$. We are forced by reasons of personal taste to leave it to the reader to determine which is meant in any given formula. To alleviate some of the pressure, we use the symbol T^* for the involution in the aforementioned C^*-algebra $\mathcal{A}_\ell(X)$. We will use the word *projection* both for an idempotent linear map, and for an orthogonal projection in an operator algebra.

An operator space is a linear subspace of $B(H)$ for a Hilbert space H. In this article all operator spaces are norm complete. Equivalently, there is an abstract characterization in terms of 'matrix norms' due to Ruan. Basic notation and facts about operator spaces may be found in [**BLM, ER4, Pau, Pis1**], for example. We recall that if X is an operator space, then so is X^*; its matrix norms come from the identification $M_n(X^*) \cong CB(X, M_n)$ (here and throughout 'CB' stands for 'completely bounded'). A *dual operator space* is one which is completely isometrically isomorphic to the operator space dual of another operator space. Any σ-weakly closed subspace of $B(H)$ is a dual operator space with predual which may be specified in terms of the predual of $B(H)$. Conversely, any dual operator space is linearly completely isometrically isomorphic and weak-$*$ homeomorphic to a σ-weakly closed subspace of some $B(H)$. We also recall from basic operator space theory that for a dual operator space $X = Y^*$, the space $CB(X)$ is canonically a dual operator space too. A bounded net $T_i \in CB(X)$ converges in the weak-$*$ topology to $T \in CB(X)$ if and only if $T_i(x) \to T(x)$ weak-$*$ in X, for every $x \in X$. We use H_c and H_r for the Hilbert column and row spaces associated to a Hilbert space H (see the texts cited above and Appendix B below).

We will use the term W^*-algebra for a C^*-algebra with predual. In view of a well-known theorem of Sakai [**Sak1**] this is 'the same as' a von Neumann algebra. For a unital Banach algebra $\mathcal{A}$ we write $\mathrm{Her}(\mathcal{A})$ for the set of Hermitian elements; these are the elements $h \in \mathcal{A}$ such that $f(h) \in \mathbb{R}$ for every state f on $\mathcal{A}$. Equivalently: $\| \exp(ith) \| \leq 1$ for all $t \in \mathbb{R}$. The Hermitians in a C^*-algebra are exactly the self-adjoint elements. We will be interested in Hermitians in the Banach algebra $B(X)$.

For the classical theory of M-ideals, the standard source is [**HWW**]. It would also be helpful if the reader was at least vaguely familiar with [**BEZ**], and had access to [**B6, BSZ**].

Acknowledgments. This memoir would not exist without the vision of Ed Effros, who originally proposed the program of generalizing the basic calculus of Banach space M-ideals to 'one-sided M-ideals' in operator spaces. We thank Larry Brown and Haskell Rosenthal for steering us toward some helpful papers, and Marius Junge and David Sherman for some helpful input during conversations on non-commutative L^p spaces and Morita equivalence.

CHAPTER 2

Preliminaries

2.1. One-Sided Multipliers

We begin by summarizing some facts about one-sided multipliers of operator spaces. For more detailed information the reader is referred to [**BLM**], or, for a quick survey addressed to a very general readership, to [**BZ**].

Let X be an operator space. Following [**B6**] §4 (see also [**WWer**]), we say that a map $T : X \to X$ is a *left multiplier* of X if there exists a linear complete isometry $\sigma : X \to B(H)$ and an operator $A \in B(H)$ such that

$$(2.1) \qquad\qquad \sigma(Tx) = A\sigma(x)$$

for all $x \in X$. In that case, we refer to (σ, A) as an *implementing pair* for T. It is easy to see that every left multiplier of X is linear. We denote by $\mathcal{M}_\ell(X)$ the set of all left multipliers of X, and for $T \in \mathcal{M}_\ell(X)$, we define the *multiplier norm* by

$$(2.2) \qquad \|T\|_{\mathcal{M}_\ell(X)} = \inf\{\|A\| : (\sigma, A) \text{ is an implementing pair for } T\}.$$

It turns out that the infimum in Equation (2.2) is always achieved. This may be seen as a consequence of the existence of the Arveson-Hamana Shilov boundary, or of the injective envelope [**Arv1, Arv2, Ham, B6, BP**]. Clearly,

$$(2.3) \qquad\qquad \|T\|_{cb} \le \|T\|_{\mathcal{M}_\ell(X)}$$

for all $T \in \mathcal{M}_\ell(X)$, so that $\mathcal{M}_\ell(X) \subset CB(X)$ as sets. In fact, strict inequality is possible in (2.3) (see [**B6**]), but this will not be of concern to us.

It can be shown that $\mathcal{M}_\ell(X)$ is a unital Banach algebra with respect to the norm $\|\cdot\|_{\mathcal{M}_\ell(X)}$, and the usual composition product. In fact, it is a unital operator algebra with respect to the operator space structure induced by the canonical isomorphisms

$$(2.4) \qquad M_n(\mathcal{M}_\ell(X)) \cong \mathcal{M}_\ell(C_n(X)) \cong \mathcal{M}_\ell(M_n(X)).$$

Whereas the definition of a left multiplier in terms of the existence of an implementing pair is extrinsic, there is an extremely useful intrinsic definition:

THEOREM 2.1 ([**BEZ**], Theorem 4.6). *Let X be an operator space and $T : X \to X$ be linear. Then T is an element of the closed unit ball of $\mathcal{M}_\ell(X)$ if and only if the map*

$$\tau_T^c : C_2(X) \to C_2(X) : \begin{bmatrix} x \\ y \end{bmatrix} \mapsto \begin{bmatrix} Tx \\ y \end{bmatrix}$$

is completely contractive.

Similar definitions and results hold for the *right multiplier algebra* $\mathcal{M}_r(X)$. Left multiplication must be replaced by right multiplication, and columns must be replaced by rows. There is a slight twist though—we regard $\mathcal{M}_r(X)$ as a subset of $CB(X)^{\mathrm{op}}$ (the opposite algebra) rather than as a subset of $CB(X)$ itself, so that the multiplication on $\mathcal{M}_r(X)$ is the reverse of the usual composition product. The reason for this is that X should be a right $\mathcal{M}_r(X)$-module. Any left multiplier on X commutes with every right multiplier on X. Thus, if $S \in \mathcal{M}_\ell(X)$ and $T \in \mathcal{M}_r(X)$, then the expression ST is unambiguous.

2.2. One-Sided Adjointable Multipliers

Let X be an operator space and $T : X \to X$ be a map. Again following [**B6**] §4, we say that T is a *left adjointable map* of X if there exists a linear complete isometry $\sigma : X \to B(H)$ and a map $S : X \to X$ such that

$$(2.5) \qquad\qquad \sigma(Tx)^*\sigma(y) = \sigma(x)^*\sigma(Sy)$$

for all $x, y \in X$. Left adjointable maps are linear and bounded. The collection of all left adjointable maps of X will be denoted $\mathcal{A}_\ell(X)$. If $T \in \mathcal{A}_\ell(X)$, then there exists a unique map $S : X \to X$ satisfying Equation (2.5). Henceforth, we will denote this map T^*. We have $T^* \in \mathcal{A}_\ell(X)$, and $(T^*)^* = T$. Every left adjointable map of X is a left multiplier of X. That is, $\mathcal{A}_\ell(X) \subset \mathcal{M}_\ell(X)$. In fact, $T \in \mathcal{A}_\ell(X)$ if and only if there exists a linear complete isometry $\sigma : X \to B(H)$ and an operator $A \in B(H)$ such that (σ, A) is an implementing pair for T and $A^*\sigma(X) \subset \sigma(X)$. Indeed, (σ, A^*) is then an implementing pair for T^*. Because of this last fact, we will often refer to left adjointable maps as *left adjointable multipliers*. For $T \in \mathcal{A}_\ell(X)$,

$$(2.6) \qquad\qquad \|T\|_{\mathcal{M}_\ell(X)} = \|T\|_{cb} = \|T\|.$$

Also, $\mathcal{A}_\ell(X)$ is a C^*-algebra with respect to the involution $\star$ and the usual composition product. In fact, $\mathcal{A}_\ell(X)$ is the 'diagonal' of the operator algebra $\mathcal{M}_\ell(X)$ (we recall that any nonselfadjoint operator algebra $\mathcal{B} \subset B(H)$ contains a canonical C^*-algebra called the diagonal, namely $\mathcal{B} \cap \mathcal{B}^*$, and the latter is well-defined independent of the representation of $\mathcal{B}$ on a Hilbert space).

There are intrinsic characterizations of certain classes of left adjointable multipliers, analogous to Theorem 2.1. Namely,

THEOREM 2.2 ([**BEZ**], Corollary 4.8 and Theorem 4.9). *Let X be an operator space and $T : X \to X$ be linear. Then*

(i) *T is a unitary element of $\mathcal{A}_\ell(X)$ if and only if $\tau_T^c : C_2(X) \to C_2(X)$ is a completely isometric surjection.*

(ii) *T is a self-adjoint element of $\mathcal{A}_\ell(X)$ if and only if τ_T^c is a Hermitian element of $CB(C_2(X))$.*

Using Theorem 2.1, one may show:

THEOREM 2.3 ([**BEZ**], Theorem 5.4 and [**B3**], Corollary 3.2). *For an operator space X, $\mathcal{M}_\ell(X^*)$ is a dual operator algebra, and $\mathcal{A}_\ell(X^*)$ is a W^*-algebra. A bounded net (T_i) in $\mathcal{M}_\ell(X^*)$ (resp. in $\mathcal{A}_\ell(X^*)$) converges weak-* to T in $\mathcal{M}_\ell(X^*)$ (resp. in $\mathcal{A}_\ell(X^*)$) if and only if $T_i \to T$ in the point-weak-* topology (that is, if and only if $T_i f \to T f$ weak-* for all $f \in X^*$).*

Furthermore, every element of $\mathcal{A}_\ell(X^*)$ is weak-* continuous ([**BEZ**], Theorem 5.5). In fact, very recently it was discovered that every element of $\mathcal{M}_\ell(X^*)$ is weak-* continuous. The interested reader is referred to [**BM**] for this, and for other complements to Theorem 2.3.

Similar definitions and results hold for the *right adjointable multiplier algebra* $\mathcal{A}_r(X)$.

The algebras $\mathcal{M}_\ell(X)$, $\mathcal{A}_\ell(X)$, $\mathcal{M}_\mathrm{r}(X)$, and $\mathcal{A}_\mathrm{r}(X)$ are invariants for the operator space X. If $\phi : X \to Y$ is a completely isometric surjection, then the map $\Phi : CB(X) \to CB(Y) : T \mapsto \phi \circ T \circ \phi^{-1}$ restricts to a completely isometric isomorphism $\mathcal{M}_\ell(X) \to \mathcal{M}_\ell(Y)$ (resp. $\mathcal{M}_\mathrm{r}(X) \to \mathcal{M}_\mathrm{r}(Y)$) and to a *-isomorphism $\mathcal{A}_\ell(X) \to \mathcal{A}_\ell(Y)$ (resp $\mathcal{A}_\mathrm{r}(X) \to \mathcal{A}_\mathrm{r}(Y)$). This can be seen in a number of ways. We will need the following result a couple of times:

LEMMA 2.4. *Let $\mathcal{A}$ be a unital Banach algebra which is also an operator space. If $T \in \mathcal{A}_\ell(\mathcal{A})_{sa}$, then $T(1) \in \mathrm{Her}(\mathcal{A})$.*

PROOF. Without loss of generality, we may assume that $0 \leq T \leq I$. Then we have
$$\|T(1) + it1\| = \|(T + itI)(1)\| \leq \|T + itI\| \leq \sqrt{1 + t^2}$$
for all $t \in \mathbb{R}$. Thus, by I.10.10 in [**BD**], we may deduce that $T(1) \in \mathrm{Her}(\mathcal{A})$. □

Finally, we remark that [**BP**] presents a very effective framework for the above multiplier algebras, in terms of the injective envelope. Since we do not essentially need this perspective, we shall not emphasize it here, although it is often useful.

2.3. One-Sided M- and L-Structure

Let X be an operator space. Following [**BEZ**], we define a *complete left M-projection* on X to be an orthogonal projection $P \in \mathcal{A}_\ell(X)$. Actually, though it is abusive to do so, we will abbreviate 'complete left M-projection' by 'left M-projection'. Hopefully the reader will agree that the savings in verbiage outweigh the possible confusion with [**BEZ**]. Being a left M-projection is equivalent to a number of other conditions, many of which are often easier to verify in practice:

THEOREM 2.5 ([**BEZ**], Proposition 3.2 and Theorem 5.1). *Let X be an operator space and $P : X \to X$ be a linear idempotent map. Then the following are equivalent:*

(i) *P is a left M-projection.*
(ii) *The map $\tau_P^c : C_2(X) \to C_2(X)$ is completely contractive. Equivalently, by Theorem 2.1, P is an element of the closed unit ball of $\mathcal{M}_\ell(X)$.*
(iii) *The map τ_P^c is a Hermitian element of $CB(C_2(X))$. Equivalently, by Theorem 2.2, $P \in \mathcal{A}_\ell(X)_{sa}$.*
(iv) *The map $\nu_P^c : X \to C_2(X) : x \mapsto \begin{bmatrix} Px \\ (\mathrm{Id} - P)x \end{bmatrix}$ is completely isometric.*
(v) *The maps ν_P^c and $\mu_P^c : C_2(X) \to X : \begin{bmatrix} x \\ y \end{bmatrix} \mapsto Px + (\mathrm{Id} - P)y$ are completely contractive.*
(vi) *There exists a completely isometric embedding $\sigma : X \hookrightarrow B(H)$, and a projection $e \in B(H)$ such that $\sigma(Px) = e\sigma(x)$ for all $x \in X$.*

A linear subspace J of X is a *right M-summand* of X if it is the range of a left M-projection P on X. The kernel of P, which equals the range of $I - P$, is also a right M-summand; it is called the *complementary right M-summand* to J. We often write this complementary summand as $\tilde{J}$, and write $X = J \oplus_{\mathrm{rM}} \tilde{J}$. Another way of stating this identity, by equivalence (iv) in the last theorem, is that there is a completely isometric embedding σ of X into $B(L, H \oplus K)$, such that $P_K \sigma(J) = P_H \sigma(\tilde{J}) = 0$.

Since left M-projections are (completely) contractive, right M-summands are automatically closed. Furthermore, since left M-projections on a dual operator space are weak-* continuous, right M-summands on such spaces are automatically weak-* closed.

A closed linear subspace J of an operator space X is a *right M-ideal* of X if its double annihilator, $J^{\perp\perp}$, is a right M-summand of X^{**}. Every right M-summand of X is a right M-ideal of X, but the converse is false.

Dual to the one-sided M-structure of an operator space X is its one-sided L-structure. A linear idempotent map $P : X \to X$ is a *right L-projection* on X if $P^* : X^* \to X^*$ is a left M-projection on X^*. It is useful, as in the case of left M-projections, to have alternative characterizations of right L-projections:

PROPOSITION 2.6 ([**BEZ**], Proposition 3.4 and Corollary 3.5). Let X be an operator space and $P : X \to X$ be a linear idempotent map. Then the following are equivalent:

(i) P is a right L-projection.
(ii) The map $\nu_P^r : X \to R_2 \hat{\otimes} X : x \mapsto \begin{bmatrix} Px & (\mathrm{Id} - P)x \end{bmatrix}$ is a complete isometry.
(iii) The maps ν_P^r and $\mu_P^r : R_2 \hat{\otimes} X \to X : \begin{bmatrix} x & y \end{bmatrix} \mapsto Px + (\mathrm{Id} - P)y$ are completely contractive.

Here $\hat{\otimes}$ is the operator space projective tensor product.

A linear subspace J of X is a *left L-summand* of X if it is the range of a right L-projection. As with right M-summands, the fact that left L-summands are closed is automatic. On the other hand, there is no need to define the concept of a *left L-ideal*—a closed linear subspace J of X is a left L-summand of X if and only if its double annihilator, $J^{\perp\perp}$, is a left L-summand of X^{**} ([**BEZ**], Proposition 3.9).

We end this section with some assorted results about the one-sided M- and L-structure of a general operator space X. These can be found in [**BEZ**] or [**Z1**]. The reader may want to take a few moments to absorb these, since they will be used freely and usually silently in the sequel.

- Let $P : X \to X$ be a bounded linear idempotent map. Then P is a left M-projection on X if and only if P^* is a right L-projection on X^*.
- A closed linear subspace J of X is a right M-ideal of X if and only if $J^\perp$ is a left L-summand of X^*. A closed linear subspace J of X is a left L-summand of X if and only if $J^\perp$ is a right M-summand of X^*.
- Every right M-summand (resp. left L-summand) is the range of a unique left M-projection (resp. right L-projection).
- If a right M-ideal J is the range of a contractive projection P, then it is in fact a right M-summand (and P is the unique left M-projection onto J). In particular, a right M-ideal which is also a dual Banach space must be a right M-summand (cf. Lemma A.2). Hence, a weak-* closed right M-ideal of X^* is a right M-summand of X^*.
- Every right M-ideal J of X is 'Hahn-Banach smooth', which is to say that each $f \in J^*$ has a unique norm-preserving extension $\tilde{f} \in X^*$. Indeed, it is easy to see that any right M-ideal J is an *HB-subspace* (see [**HWW**, p. 44]), and therefore enjoys the properties of such spaces. In particular J^* may be identified with a closed linear subspace of X^*, none other than the complementary left L-summand of $J^\perp$. That is, $X^* = J^\perp \oplus_{\ell\mathrm{L}} J^*$.

Another useful sufficient condition for a right M-ideal to be a right M-summand comes from a fact from [**GKS**]. Namely, those authors define a class of subspaces of a Banach space called *u-ideals*. It is easy to see that every one-sided M-ideal is a *u*-ideal. On the other hand, from [**GKS**, p.14] we know that any *u*-ideal not containing a linear topologically isomorphic copy of c_0 is a *u-summand*. It is easy to see that any *u*-summand is contractively complemented. From this and the second last 'bullet' above, we deduce that:

- Any one-sided M-ideal not containing a linear topologically isomorphic copy of c_0 is a one-sided M-summand.

2.4. The One-Sided Cunningham Algebra

Let X be an operator space. Following [**Z1**], we define the *right Cunningham algebra* of X to be the closed linear span in $CB(X)^{\mathrm{op}}$ of the right L-projections on X:

$$\mathcal{C}_{\mathrm{r}}(X) = \overline{\operatorname{span}}\{P : P \text{ is a right } L\text{-projection on } X\}.$$

The closure here is in the associated norm, namely the 'cb' norm on $CB(X)^{\mathrm{op}}$. For $T \in \mathcal{C}_{\mathrm{r}}(X)$, one has that

$$(2.7) \qquad\qquad\qquad\qquad \|T\|_{cb} = \|T\|.$$

This follows from the fact that for any T in the span of the right L-projections on X, T^* is in the span of the left M-projections on X^*, and therefore is an element of $\mathcal{A}_\ell(X^*)$, so that

$$\|T\|_{cb} = \|T^*\|_{cb} = \|T^*\| = \|T\|$$

by Equation (2.6). Clearly, the isometric homomorphism $CB(X)^{\mathrm{op}} \to CB(X^*)$: $T \mapsto T^*$ restricts to a linear isometry $\rho : \mathcal{C}_{\mathrm{r}}(X) \to \mathcal{A}_\ell(X^*)$. Because $\mathcal{A}_\ell(X^*)$ is a W^*-algebra (Theorem 2.3), it is the norm-closed linear span of its projections (the left M-projections on X^*). Since every left M-projection on X^* is weak-* continuous (cf. Section 2.3), and thus the adjoint of a right L-projection on X, ρ is surjective. It follows that $\mathcal{C}_{\mathrm{r}}(X)$ is a (closed) subalgebra of $CB(X)^{\mathrm{op}}$, and that the contractive projections in $\mathcal{C}_{\mathrm{r}}(X)$ are exactly the right L-projections on X. Endowed with the involution defined by

$$T^\star = \rho^{-1}(\rho(T)^\star),$$

$\mathcal{C}_{\mathrm{r}}(X)$ becomes a C^*-algebra. Obviously, ρ is then a *-isomorphism. By Sakai's theorem [**Sak1**], $\mathcal{C}_{\mathrm{r}}(X)$ is actually a W^*-algebra with unique predual, and ρ is a weak-* homeomorphism. A bounded net (T_i) in $\mathcal{C}_{\mathrm{r}}(X)$ converges weak-* to $T \in \mathcal{C}_{\mathrm{r}}(X)$ if and only if $T_i(x) \to T(x)$ weakly for all $x \in X$.

Similar definitions and results hold for the *left Cunningham algebra*, $\mathcal{C}_\ell(X)$.

CHAPTER 3

Spatial Action

We have seen that to each operator space X, there are associated natural C^*-algebras $\mathcal{A}_\ell(X)$ and $\mathcal{A}_r(X)$, and W^*-algebras $\mathcal{C}_r(X)$ and $\mathcal{C}_\ell(X)$. For dual operator spaces, the C^*-algebras are in fact W^*-algebras. These operator algebras are not concretely given, which is to say that they do not appear acting on a Hilbert space. On the other hand, they are not totally 'space-free': they do act on X. So one can hope to employ spatial intuition and arguments. Of course, one must exercise caution in doing so—X is typically far from being a Hilbert space. In this chapter we describe how projections and partial isometries in our abstract operator algebras manipulate the underlying space X.

3.1. Projections

First we investigate the partial ordering of projections. Recall that for projections E and F in an abstract C^*-algebra $\mathcal{A}$, $E \leq F$ if and only if $EF = FE = E$. If $\mathcal{A}$ is a $*$-subalgebra of $B(H)$, then $E \leq F$ if and only if $E(H) \subset F(H)$. It is reassuring that the analogous result holds in our setting.

PROPOSITION 3.1. Let X be an operator space, and P and Q be left M-projections (resp. right L-projections) on X. Then $P \leq Q$ if and only if $P(X) \subset Q(X)$.

PROOF. Suppose $P \circ Q = Q \circ P = P$. Then $P(X) \subset Q(X)$. Conversely, suppose that $P(X) \subset Q(X)$. Then $Q \circ P = P$. But then $P \circ Q = (Q \circ P)^\star = P^\star = P$. $\square$

Next we investigate the lattice of projections. Recall that for an abstract W^*-algebra $\mathcal{A}$, the projections in $\mathcal{A}$ form a complete lattice with respect to the partial ordering of projections. If $\mathcal{A} \subset B(H)$ and $\{E_i : i \in I\}$ is a family of projections in $\mathcal{A}$, then $\bigwedge_{i \in I} E_i$ is the projection onto $\bigcap_{i \in I} E_i(H)$ and $\bigvee_{i \in I} E_i$ is the projection onto $\overline{\operatorname{span}}\{\bigcup_{i \in I} E_i(H)\}$. To determine what happens in our setting, we need abstract (space-free) formulas for the meet and join of projections in a W^*-algebra. Such formulas were first considered by von Neumann:

LEMMA 3.2 ([vN]). Let $\mathcal{A} \subset B(H)$ be a W^*-algebra and $E, F \in \mathcal{A}$ be projections. Then
$$E \wedge F = \text{wk*-}\lim_{n \to \infty} (EF)^n$$
and
$$E \vee F = \text{wk*-}\lim_{n \to \infty} I - (I - E - F + EF)^n,$$
where $I \in B(H)$ is the identity.

We now apply Lemma 3.2 to our situation.

PROPOSITION 3.3. Let X be a dual operator space, and P and Q be left M-projections on X. Then

(i) $(P \wedge Q)(X) = P(X) \cap Q(X)$.

(ii) $(P \vee Q)(X) = \overline{P(X) + Q(X)}^{\text{wk}^*}$.

PROOF. (i) By Proposition 3.1, $(P \wedge Q)(X) \subset P(X) \cap Q(X)$. Conversely, suppose that $x \in P(X) \cap Q(X)$. Then by Lemma 3.2 and Theorem 2.3, we have that

$$(P \wedge Q)(x) = \text{wk}^*\text{-}\lim_{n \to \infty} (PQ)^n(x) = x,$$

which says that $x \in (P \wedge Q)(X)$.

(ii) Again by Proposition 3.1, $P(X) + Q(X) \subset (P \vee Q)(X)$. Since $(P \vee Q)(X)$ is weak-* closed,

$$\overline{P(X) + Q(X)}^{\text{wk}^*} \subset (P \vee Q)(X).$$

Conversely, suppose that $x \in (P \vee Q)(X)$. Then by Lemma 3.2 and Theorem 2.3 again, we have that

$$x = (P \vee Q)(x) = \text{wk}^*\text{-}\lim_{n \to \infty} x - (I - P - Q + PQ)^n(x).$$

Since $x - (I - P - Q + PQ)^n(x) = P(X) + Q(X)$ for all $n \in \mathbb{N}$, it follows that $x \in \overline{P(X) + Q(X)}^{\text{wk}^*}$. $\square$

COROLLARY 3.4. Let X be an operator space, and P and Q be right L-projections on X. Then

(i) $(P \wedge Q)(X) = P(X) \cap Q(X)$.

(ii) $(P \vee Q)(X) = \overline{P(X) + Q(X)}$.

The proof is the same as that of Proposition 3.3, modulo the obvious modifications. Alternatively, one can deduce Corollary 3.4 from Proposition 3.3 by exploiting the duality between one-sided M-structure and one-sided L-structure.

The passage from the finite to the infinite is straightforward.

PROPOSITION 3.5.   (i) If X is a dual operator space and $\{P_i : i \in I\}$ is a family of left M-projections on X, then $\left(\bigwedge_{i \in I} P_i\right)(X) = \cap_{i \in I} P_i(X)$ and $\left(\bigvee_{i \in I} P_i\right)(X) = \overline{\text{span}}^{\text{wk}^*}\left\{\bigcup_{i \in I} P_i(X)\right\}$.

(ii) If Y is an operator space and $\{Q_j : j \in J\}$ is a family of right L-projections on Y, then $\left(\bigwedge_{j \in J} Q_j\right)(Y) = \cap_{j \in J} Q_j(Y)$ and $\left(\bigvee_{j \in J} Q_j\right)(Y) = \overline{\text{span}}\left\{\bigcup_{j \in J} Q_j(Y)\right\}$.

PROOF. We will prove half of each of the two assertions. The reader will have no trouble supplying the missing arguments.

(i) For each finite subset F of I, define $P_F = \bigwedge_{i \in F} P_i$. By Proposition 3.3, $P_F(X) = \cap_{i \in F} P_i(X)$. Also, $P_F \to P \equiv \bigwedge_{i \in I} P_i$ weak-*. Since $P \leq P_i$ for all $i \in I$, $P(X) \subset P_i(X)$ for all $i \in I$ (Proposition 3.1). Therefore, $P(X) \subset \cap_{i \in I} P_i(X)$. On the other hand, if $x \in \cap_{i \in I} P_i(X)$, then $P(x) = \text{wk}^*\text{-}\lim_F P_F(x) = x$, so that $x \in P(X)$. Hence, $\cap_{i \in I} P_i(X) \subset P(X)$.

(ii) For each finite subset F of J, define $Q_F = \bigvee_{j \in F} Q_j$. By Corollary 3.4, $Q_F(X) = \overline{\text{span}}\left\{\bigcup_{j \in F} Q_j(X)\right\}$. Also, $Q_F \to Q \equiv \bigvee_{j \in J} Q_j$ weak-*. Since $Q_j \leq Q$ for all $j \in J$, $Q_j(X) \subset Q(X)$ for all $j \in J$, which implies that $\overline{\text{span}}\left\{\bigcup_{j \in J} Q_j(X)\right\} \subset Q(X)$. On the other hand, if $x \in Q(X)$, then $Q_F(x) \to x$ weakly. Thus, $x \in$

$\overline{\operatorname{span}}^{\mathrm{wk}}\left\{\bigcup_{j\in J}Q_j(X)\right\}=\overline{\operatorname{span}}\left\{\bigcup_{j\in J}Q_j(X)\right\}$. Since the choice of x was arbitrary, $Q(X)\subset\overline{\operatorname{span}}\left\{\bigcup_{j\in J}Q_j(X)\right\}$. $\qquad\square$

COROLLARY 3.6.   (i) If X is an operator space, then there is a lattice isomorphism between the complete lattice of right L-projections on X, and the lattice of left L-summands of X.

(ii) Let X be a dual operator space, and define the 'meet' of a family of subspaces of X to be their intersection, and the 'join' to be the weak-* closure of their span. Then there is a lattice isomorphism between the complete lattice of left M-projections on X, and the lattice of right M-summands of X.

3.2. Partial Isometries

Next we investigate partial isometries W in $\mathcal{A}_\ell(X)$. Recall that an element W of a C^*-algebra $\mathcal{A}$ is a partial isometry if W^*W is a projection. We assume familiarity with the basic facts about partial isometries on Hilbert spaces (see [**KR2**]).

PROPOSITION 3.7. Let X be an operator space and $W\in\mathcal{A}_\ell(X)$ be a partial isometry. Then $P=W^\star W$ and $Q=WW^\star$ are left M-projections on X. One has that

$$\operatorname{ran}(P)=\operatorname{ran}(W^\star),\ \ker(P)=\ker(W),\ \operatorname{ran}(Q)=\operatorname{ran}(W),\ \text{and}\ \ker(Q)=\ker(W^\star).$$

In particular, $\operatorname{ran}(W)$ and $\operatorname{ran}(W^\star)$ are closed and

$$X=\operatorname{ran}(W^\star)\oplus_{\mathrm{rM}}\ker(W)=\operatorname{ran}(W)\oplus_{\mathrm{rM}}\ker(W^\star).$$

Finally, W maps $\operatorname{ran}(W^\star)$ completely isometrically onto $\operatorname{ran}(W)$.

PROOF. Clearly P and Q are left M-projections on X. We have that

$$\operatorname{ran}(P)=\operatorname{ran}(W^*W)\subset\operatorname{ran}(W^\star)=\operatorname{ran}(W^\star WW^\star)=\operatorname{ran}(PW^\star)\subset\operatorname{ran}(P).$$

Likewise, $\operatorname{ran}(Q)=\operatorname{ran}(W^\star)$. Clearly, $Wx=0$ implies $Px=W^\star Wx=0$. Conversely, $Px=0$ implies $Wx=WW^\star Wx=WPx=0$. Hence, $\ker(P)=\ker(W)$. Likewise, $\ker(Q)=\ker(W^\star)$. Clearly, W is a linear bijection between $\operatorname{ran}(W^\star)$ and $\operatorname{ran}(W)$ (with inverse $W^\star$). Since both W and $W^\star$ are completely contractive, the last assertion follows. $\qquad\square$

The same result is true if we replace $\mathcal{A}_\ell(X)$ by $\mathcal{C}_{\mathrm{r}}(X)$, modulo the obvious modifications.

3.3. Murray-Von Neumann Equivalence

Let X be an operator space and let P and Q be left M-projections on X. We say that P and Q are *(left) Murray-von Neumann equivalent* (or simply *equivalent*), and write $P\sim Q$, if there exists a partial isometry $V\in\mathcal{A}_\ell(X)$ such that $V^\star V=P$ and $VV^\star=Q$ (i.e. if they are equivalent in $\mathcal{A}_\ell(X)$). Suppose $P\sim Q$ (via V), and let J and K be the right M-summands corresponding to P and Q, respectively. Denote by $\tilde{J}$ the complementary right M-summand of J, and by $\check{K}$ the complementary right M-summand of K. By Proposition 3.7, we have that $\ker(V)=\tilde{J}$ and V maps J completely isometrically onto K. If, further, X is a dual operator space, then as we stated after Theorem 2.3 and Theorem 2.5, V and $V^\star$ are weak-* continuous, and J and K are weak-* closed. Thus J is also weak-* homeomorphic to K.

It is clear that if $\mathcal{A}_\ell(X)$ is commutative, then Murray-von Neumann equivalent left M-projections on X must be equal. Conversely, a characterization of noncommutativity in von Neumann algebras is the existence of two equivalent mutually orthogonal nonzero projections. Furthermore, it follows easily from the well-known 'comparison theorem' for von Neumann algebra projections that if P, Q are two non-commuting left M-projections on a dual operator space X, then P and Q have nonzero subprojections which are Murray-von Neumann equivalent.

3.4. Inner Products on Operator Spaces

A perspective emphasized in [**B6**] is that a general operator space X may be studied using the framework of Hilbert C^*-modules (see e.g. [**Lan**]). Indeed, every operator space X has canonical C^*-algebra-valued inner products, which we will call the *Shilov inner products*. These may be viewed as the natural inner products on the Hamana triple envelope of X, or may be described in terms of the injective envelope. Since the reader may not be familiar with these notions, we will not go into much detail. Suffice it to say that a 'left Shilov inner product' $\langle \cdot, \cdot \rangle$ is a map from $X \times X$ into a C^*-algebra, which is linear in the second variable and conjugate linear in the first variable. It is explicitly described at the top of p. 307 in [**B6**], and it has the following property: $T \in \mathcal{A}_\ell(X)$ if and only if there is a map $S \, (= T^\star)$ on X with

$$(3.1) \qquad \langle Tx, y \rangle = \langle x, Sy \rangle$$

for all $x, y \in X$. Also $\|\langle x, x \rangle\| = \|x\|^2$ for $x \in X$. In fact we shall be more concerned here with the associated notion of orthogonality with respect to the inner product: we write $x \curlywedge y$ if and only if $\langle x, y \rangle = 0$. This has a simple reformulation: from the universal property of the Hamana triple envelope it is easy to see that $x \curlywedge y$ if and only if there exists a completely isometric embedding σ of X into a C^*-algebra with

$$(3.2) \qquad \sigma(x)^* \sigma(y) = 0.$$

Similarly, for subsets $B, C \subset X$, we write $B \curlywedge C$ if $x \curlywedge y$ for all $x \in B, y \in C$. It can be easily shown that this is equivalent to the existence of a map σ as above such that (3.2) holds for every choice of $x \in B, y \in C$. It follows that if $x \curlywedge y$ in an operator space X, and if $x, y \in Y \subset X$, then $x \curlywedge y$ in Y. (The converse of this is false.)

In [**BEZ**, Theorem 5.1] we showed that a linear idempotent map P on an operator space X is a left M-projection if and only if $P(X) \curlywedge (I - P)(X)$. It follows that if J, K are linear subspaces of X with $J + K = X$ and $J \curlywedge K$, then J is a right M-summand of X and K is the complementary right M-summand. More generally, if $T \in \mathcal{A}_\ell(X)$, then $\ker(T^\star) \curlywedge \overline{\operatorname{ran}(T)}$.

3.5. Polar Decomposition

In this section we determine when a left adjointable multiplier has a 'nice' polar decomposition. We shall see that on a general operator space X, not every $T \in \mathcal{A}_\ell(X)$ is 'polar decomposable'. We give necessary and sufficient conditions for this to occur (Theorem 3.8). One important instance when these conditions are met is when T has closed range (Theorem 3.9), and we draw a number of conclusions from this. Of course, if X is a dual operator space, then T automatically has a nice polar decomposition (Theorem 3.13). Again, this has favorable consequences.

As the reader shall see, much of this section is a good illustration of the remark in the previous section that a general operator space X may be studied using the framework of Hilbert C^*-modules.

For $T \in \mathcal{A}_\ell(X)$, we write $|T|$ for the square root of $T^\star T$ in $\mathcal{A}_\ell(X)$. Note that if $\langle \cdot, \cdot \rangle$ is the 'left Shilov inner product' on X (see Section 3.4), then

$$\langle Tx, Tx \rangle = \langle T^\star Tx, x \rangle = \langle |T|x, |T|x \rangle,$$

for $x \in X$, using (3.1). This together with the fact that $\|\langle x, x \rangle\| = \|x\|^2$ shows that $\ker(T) = \ker(|T|)$.

As is remarked in passing in [**B6**], Appendix B, the entire discussion in Section 15.3 of [**Weg**] on the polar decomposition of adjointable operators on a Hilbert C^*-module carries over verbatim to the setting of left adjointable multipliers of an operator space. Using arguments identical to those found in [**Weg**, Section 15.3], one obtains the following two theorems:

THEOREM 3.8. *Let X be an operator space and $T \in \mathcal{A}_\ell(X)$. Then the following are equivalent:*

(i) $T = W|T|$ *for some partial isometry $W \in \mathcal{A}_\ell(X)$ such that* $\mathrm{ran}(W^\star) = \overline{\mathrm{ran}(|T|)}$ *and* $\mathrm{ran}(W) = \overline{\mathrm{ran}(T)}$.

(ii) $X = \overline{\mathrm{ran}(|T|)} \oplus_{\mathrm{rM}} \ker(|T|) = \overline{\mathrm{ran}(T)} \oplus_{\mathrm{rM}} \ker(T^\star)$.

(iii) $\overline{\mathrm{ran}(|T|)}$ *and* $\overline{\mathrm{ran}(T)}$ *are right M-summands of X.*

Of course, it follows too that if (i)-(iii) of the last theorem hold, then $\ker(|T|)$ and $\ker(T^\star)$ are right M-summands of X.

THEOREM 3.9. *Suppose that T is a left adjointable multiplier of an operator space X. If $\mathrm{ran}(T)$ is closed in X, then $T^\star$ and $|T|$ have closed ranges as well. Also, $\ker(T), \ker(T^\star), \mathrm{ran}(T)$, and $\mathrm{ran}(T^\star)$ are right M-summands of X. We have:*

$$X = \ker(T) \oplus_{\mathrm{rM}} \mathrm{ran}(T^\star) = \ker(T^\star) \oplus_{\mathrm{rM}} \mathrm{ran}(T) \ .$$

Finally, T has a polar decomposition $T = V|T|$ for a left adjointable partial isometry V satisfying $\ker(V) = \ker(T)$, $\ker(V^\star) = \ker(T^\star)$, $\mathrm{ran}(V) = \mathrm{ran}(T)$, *and* $\mathrm{ran}(V^\star) = \mathrm{ran}(T^\star)$.

Using Theorem 3.9, one may draw a number of important conclusions.

COROLLARY 3.10. *A closed linear subspace J of an operator space X is a right M-summand if and only if it is the range of some $T \in \mathcal{A}_\ell(X)$.*

Interestingly, for a left adjointable multiplier T on a dual operator space X, this says that $\mathrm{ran}(T)$ is closed if and only if $\mathrm{ran}(T)$ is weak-* closed.

COROLLARY 3.11. *If $T \in \mathcal{A}_\ell(X)$ for an operator space X, then the following are equivalent:*

(i) T *is a bijection,*

(ii) T *is invertible in $\mathcal{A}_\ell(X)$,*

(iii) T *and $T^\star$ are bounded away from zero. That is, there exists $\epsilon > 0$ such that $\|Tx\| \geq \epsilon \|x\|$ for all $x \in X$, and similarly for $T^\star$.*

In this case we have $(T^{-1})^\star = (T^\star)^{-1}$.

PROOF. If T is bijective, then by the displayed equation in 3.9, we have $\mathrm{ran}(T^\star) = X$ and $\ker(T^\star) = \{0\}$. So $T^\star$ is bijective. If $x, y \in X$, then

$$\langle T^{-1}x, y \rangle = \langle T^{-1}x, T^\star((T^\star)^{-1}(y)) \rangle = \langle x, (T^\star)^{-1}(y) \rangle \ ,$$

where $\langle \cdot, \cdot \rangle$ is the 'inner product' discussed in 3.4. This gives (ii) as well as the final formula. That (ii) $\Rightarrow$ (iii) is clear. Finally, given (iii), we see by the displayed equation in 3.9 that T is onto, and we obtain (i). $\qquad\square$

The following is related to a result of Lance for Hilbert C^*-modules [**Lan**, Theorem 3.5]:

COROLLARY 3.12. *Let X be an operator space. A surjective left adjointable isometry $T : X \to X$ is a unitary in $\mathcal{A}_\ell(X)$.*

PROOF. By the last result, T^{-1} is left adjointable. Since T and T^{-1} are elements of norm one in the C^*-algebra $\mathcal{A}_\ell(X)$, and since their product is the identity in either order, they must be unitary. $\qquad\square$

Now suppose that X is a dual operator space. Then $\mathcal{A}_\ell(X)$ is a W^*-algebra, which implies that every $T \in \mathcal{A}_\ell(X)$ satisfies $T = W|T|$, where $W \in \mathcal{A}_\ell(X)$ is a partial isometry such that $W^\star W$ is the left support of $|T|$ (the smallest projection $P \in \mathcal{A}_\ell(X)$ such that $P|T| = |T|$) and $WW^\star$ is the left support of T. This simple observation yields the following result.

THEOREM 3.13. *Let X be a dual operator space and $T \in \mathcal{A}_\ell(X)$. Then $T = W|T|$, where $W \in \mathcal{A}_\ell(X)$ is a partial isometry such that $\mathrm{ran}(W^\star) = \overline{\mathrm{ran}(|T|)}^{\,\mathrm{wk}^*}$ and $\mathrm{ran}(W) = \overline{\mathrm{ran}(T)}^{\,\mathrm{wk}^*}$. Furthermore, $\ker(W) = \ker(|T|)$ and $\ker(W^\star) = \ker(T^\star)$. In particular, $X = \overline{\mathrm{ran}(T)}^{\,\mathrm{wk}^*} \oplus_{\mathrm{rM}} \ker(T^\star)$.*

PROOF. Let $W \in \mathcal{A}_\ell(X)$ be the partial isometry described in the paragraph preceding the proposition. We have that

$$\mathrm{ran}(|T|) = \mathrm{ran}(W^\star T) = \mathrm{ran}(W^\star W W^\star T) \subset \mathrm{ran}(W^\star W).$$

Since $\mathrm{ran}(W^\star W)$ is weak-* closed, $\overline{\mathrm{ran}(|T|)}^{\,\mathrm{wk}^*} \subset \mathrm{ran}(W^\star W)$. On the other hand, we have by Lemma 5.1.5 of [**KR1**] that

$$W^\star W = \mathrm{wk}^*\text{-}\lim_{n\to\infty} |T|^{1/n}.$$

By spectral theory (i.e. polynomial approximation via the functional calculus) it is easy to see that $\mathrm{ran}(|T|^{1/n}) \subset \overline{\mathrm{ran}(|T|)}$ for all $n \in \mathbb{N}$. Hence $\mathrm{ran}(W^\star W) \subset \overline{\mathrm{ran}(|T|)}^{\,\mathrm{wk}^*}$. Now suppose that $|T|x = 0$. Then $|T|^{1/n}x = 0$ for all $n \in \mathbb{N}$, which implies that $W^\star W x = 0$. Conversely, suppose that $W^\star W x = 0$. Then $W x = 0$, which implies that $|T|x = T^\star W x = 0$. Hence, $\ker(W^\star W) = \ker(|T|)$. By Proposition 3.7, $\mathrm{ran}(W^\star) = \overline{\mathrm{ran}(|T|)}^{\,\mathrm{wk}^*}$ and $\ker(W) = \ker(|T|)$. This is half of what we need to prove. Proceeding to the other half, we have that

$$\mathrm{ran}(T) = \mathrm{ran}(W|T|) = \mathrm{ran}(W W^\star W|T|) \subset \mathrm{ran}(WW^\star),$$

from which it follows that $\overline{\mathrm{ran}(T)}^{\,\mathrm{wk}^*} \subset \mathrm{ran}(WW^\star)$. On the other hand,

$$WW^\star = W(W^\star W)W^\star = \mathrm{wk}^*\text{-}\lim_{n\to\infty} W|T|^{1/n}W^\star.$$

Since $\mathrm{ran}(|T|^{1/n}) \subset \overline{\mathrm{ran}(|T|)}$, we have $\mathrm{ran}(W|T|^{1/n}) \subset \overline{\mathrm{ran}(T)}$, for all $n \in \mathbb{N}$. Thus $\mathrm{ran}(WW^\star) \subset \overline{\mathrm{ran}(T)}^{\,\mathrm{wk}^*}$. Now suppose that $T^\star x = 0$. Then $|T|W^\star x = 0$, which implies that $|T|^{1/n}W^\star x = 0$ for all $n \in \mathbb{N}$. Therefore, $WW^\star x = 0$. Conversely, suppose that $WW^\star x = 0$. Then $W^\star x = 0$. Thus, $T^\star x = |T|W^\star x = 0$.

Hence, $\ker(WW^\star) = \ker(T^\star)$. By Proposition 3.7 again, $\operatorname{ran}(W) = \overline{\operatorname{ran}(T)}^{\mathrm{wk}^*}$ and $\ker(W^\star) = \ker(T^\star)$. $\square$

COROLLARY 3.14. *Let J be a linear subspace of a dual operator space X. Then the following are equivalent:*

(i) *J is a right M-summand of X.*
(ii) *$J = \overline{\operatorname{ran}(T)}^{\mathrm{wk}^*}$ for some $T \in \mathcal{A}_\ell(X)$.*
(iii) *$J = \ker(T)$ for some $T \in \mathcal{A}_\ell(X)$.*

COROLLARY 3.15. *Let X be an operator space and $T \in \mathcal{A}_\ell(X)$. Then $\overline{\operatorname{ran}(T)}$ is a right M-ideal of X.*

PROOF. By basic functional analysis we have
$$\overline{\operatorname{ran}(T)}^{\perp\perp} = \ker(T^*)^\perp = ({}^\perp\operatorname{ran}(T^{**}))^\perp = \overline{\operatorname{ran}(T^{**})}^{\mathrm{wk}^*}.$$
We shall see in Section 5.3 that $T^{**} \in \mathcal{A}_\ell(X^{**})$. Thus the result follows from Corollary 3.14. $\square$

COROLLARY 3.16. *Let X be an operator space. If X has no nontrivial right M-ideals, then $\mathcal{A}_\ell(X) \cong \mathbb{C}$.*

PROOF. Let $0 \neq T \in \mathcal{A}_\ell(X)$. By Corollary 3.15, $\overline{\operatorname{ran}(T)} = X$, since the left-hand side is a nonzero right M-ideal of X. Thus, if $S \in \mathcal{A}_\ell(X)$ and $ST = 0$, then $S = 0$. Consequently, if $A, B \in \mathcal{A}_\ell(X)$ and $AB = 0$, then either $A = 0$ or $B = 0$.

Now let $0 \neq T \in \mathcal{A}_\ell(X)_{sa}$. Assume that $\lambda_1, \lambda_2 \in \mathbb{R}$ are distinct elements of $\sigma(T)$, the spectrum of T. There exist $f_1, f_2 \in C(\sigma(T))$ such that $f_i(\lambda_i) = 1$, $i = 1, 2$, and $f_1 f_2 = 0$. But then $T_1 = f_1(T)$ and $T_2 = f_2(T)$ are nonzero elements of $\mathcal{A}_\ell(X)$ such that $T_1 T_2 = 0$, a contradiction. It follows that $\sigma(T) = \{\lambda\}$ for some $\lambda \in \mathbb{R}$, which implies that $T = \lambda I$. Thus $\mathcal{A}_\ell(X) \cong \mathbb{C}$. $\square$

The last result, like many others in our paper, has interesting consequences when applied to particular special cases (for example if X is a C^*-module, or a nonselfadjoint operator algebra). We often will not take the time to write down such ensuing corollaries.

Dual to Theorem 3.13 we have the following result:

COROLLARY 3.17. *Let X be an operator space and $T \in \mathcal{C}_\mathrm{r}(X)$. Then $T = W \circ (T^\star \circ T)^{1/2}$, where $W \in \mathcal{C}_\mathrm{r}(X)$ is a partial isometry such*
$$\operatorname{ran}(W^\star) = \overline{\operatorname{ran}(T^\star)}, \operatorname{ran}(W) = \overline{\operatorname{ran}(T)}, \ker(W) = \ker(T), \text{ and } \ker(W^\star) = \ker(T^\star).$$
In particular, $X = \overline{\operatorname{ran}(T)} \oplus_{\ell L} \ker(T^\star)$.

The proof is essentially the same as that of Theorem 3.13. Note that $T = W \circ (T^\star \circ T)^{1/2}$ is not the same as $T = W|T|$, because of our convention that $\mathcal{C}_\mathrm{r}(X) \subset CB(X)^{\mathrm{op}}$ (see Section 2.4). Indeed, the correct formula is $T = |T^\star|W$.

COROLLARY 3.18. *Let J be a linear subspace of an operator space X. Then the following are equivalent:*

(i) *J is a left L-summand of X.*
(ii) *$J = \overline{\operatorname{ran}(T)}$ for some $T \in \mathcal{C}_\mathrm{r}(X)$.*
(iii) *$J = \ker(T)$ for some $T \in \mathcal{C}_\mathrm{r}(X)$.*

Remarks. 1) Corollaries 3.10, 3.14, and 3.15 are not true in general if $T \in \mathcal{M}_\ell(X)$.

2) We shall see later (see the last remark in Section 5.3) that left adjointable multipliers on a general operator space always have a 'pseudo polar decomposition'.

3) In view of relations such as those in Corollaries 3.14, 3.15 and 3.18, it is tempting to believe that the kernel of a map $T \in \mathcal{A}_\ell(X)$ is a right M-ideal of X. But this is not true. For a counterexample, let $T = P + Q$ for two left M-projections P, Q on X. If $Tx = 0$, then

$$\langle Px + Qx, x \rangle = \langle Px, Px \rangle + \langle Qx, Qx \rangle = 0,$$

where $\langle \cdot, \cdot \rangle$ is the inner product in Section 3.4. This implies that $Px = Qx = 0$. Thus $\ker(P + Q) = \ker(P) \cap \ker(Q)$. However, we shall see in Example 5.28 that the intersection of right M-summands need not be a right M-ideal.

4) In connection to 3.15, it is not true that every right M-ideal is the closure of the range of a left adjointable multiplier. For example, $K(H)$ is a right M-ideal of $\mathrm{MIN}(B(H))$, but $\mathcal{A}_\ell(\mathrm{MIN}(B(H))) = \mathbb{C}\,\mathrm{Id}_{B(H)}$ (see Section 4.2 and Proposition 7.6). This also shows that the converse of Corollary 3.16 is false.

CHAPTER 4

Examples

Many examples of one-sided M-structure of particular operator spaces are listed in our previous papers [**BEZ, BSZ**]. We will need to restate a few main facts here, omitting proofs given in those papers to avoid repetition. Because this chapter is devoted to *examples*, we will feel free to quote from results proved only later in this memoir.

4.1. Two-Dimensional Operator Spaces

In this section we consider the one-sided M-structure of two-dimensional operator spaces. First, however, we give a result for finite-dimensional operator spaces.

PROPOSITION 4.1. Let X be a n-dimensional operator space. Then $\mathcal{A}_\ell(X)$ is *-isomorphic to a unital *-subalgebra of M_m, for some $m \leq n$.

PROOF. Since $\mathcal{A}_\ell(X) \subset B(X)$, $\mathcal{A}_\ell(X)$ is finite-dimensional, and is therefore *-isomorphic to $M_{n_1} \oplus M_{n_2} \oplus ... \oplus M_{n_k}$ for some positive integers $n_1, n_2, ..., n_k$. Since $\dim(X) = n$, $\mathcal{A}_\ell(X)$ has at most n mutually orthogonal nonzero projections. Hence, $n_1 + n_2 + ... + n_k \leq n$. □

Remark. Since $\mathcal{M}_\ell(X) \subset B(X)$, it is clear that $\mathcal{M}_\ell(X)$ is also finite dimensional for an n-dimensional operator space X. However, there may exist no finite m with $\mathcal{M}_\ell(X) \subset M_m$ completely isometrically. This may be seen via the following well-known trick: take a finite-dimensional operator space $Y \subset B(H)$, and let X be the unital operator algebra in $M_2(B(H))$ with 'Y as the $1-2$ corner', and 'scalars on the diagonal'. Then $Y \subset X = \mathcal{M}_\ell(X)$ completely isometrically. If Y is not an exact operator space [**Pis1**], then the assertion follows.

From this point forward in this subsection, X will be a two-dimensional operator space, so that the nontrivial right M-summands correspond to certain unit vectors $x \in X$. We remark in passing that if x is a unit vector in a general operator space X, and if y is a unit vector such that $y \curlywedge x$, then by the last remark in Section 3.4 one may easily see that $J_0 = \mathrm{span}\{x\}$ is a right M-summand in $X_0 = \mathrm{span}\{x, y\}$. Thus, in some sense, the 2-dimensional case captures at least some of the geometry of the general case. It would be nice to have a tidy geometric characterization of right M-summands of two-dimensional operator spaces, however this still eludes us. To appreciate some of the complications, note that surprisingly even ℓ_2^p (with some operator space structure) can have nontrivial right M-summands (see Example 5.44). We can at least say the following, which rules out the existence of right M-summands in some classical spaces:

PROPOSITION 4.2. If X is a two-dimensional operator space with a nontrivial right M-summand, then

(i) X is completely isometric to the span in $B(H)$ of two operators S and T of norm one such that $S^*T = 0$.

(ii) There is a completely contractive surjection $\rho : C_2 \to X$ taking e_1 and e_2 to unit vectors in the complementary summands.

PROOF. Let x be a unit vector in the nontrivial right M-summand J of X, and y be a unit vector in the complementary summand. Let $P \in \mathcal{A}_\ell(X)$ be the left M-projection onto J. Then there exists a complete isometry $\sigma : X \to B(H)$ such that $\sigma(Pv)^*\sigma(w) = \sigma(v)^*\sigma(Pw)$ for all $v, w \in X$. Let $S = \sigma(x)$ and $T = \sigma(y)$. Then

$$S^*T = \sigma(x)^*\sigma(y) = \sigma(Px)^*\sigma(y) = \sigma(x)^*\sigma(Py) = 0.$$

This proves (i). Now define $\rho : C_2 \to \operatorname{span}\{S, T\}$ by

$$\rho\left(\begin{bmatrix} a \\ b \end{bmatrix}\right) = aS + bT$$

for all $a, b \in \mathbb{C}$. For $A, B \in M_n$, one has that

$$
\begin{aligned}
\|A \otimes S + B \otimes T\|^2 &= \|A^*A \otimes S^*S + A^*B \otimes S^*T + B^*A \otimes T^*S + B^*B \otimes T^*T\| \\
&= \|A^*A \otimes S^*S + B^*B \otimes T^*T\| \leq \|(A^*A + B^*B) \otimes I\| \\
&= \|A^*A + B^*B\| = \left\|\begin{bmatrix} A \\ B \end{bmatrix}\right\|^2.
\end{aligned}
$$

This proves (ii). $\square$

Of course, the converse of (i) above is true as well: if X is completely isometric to the span in $B(H)$ of two operators S and T of norm one such that $S^*T = 0$, then X is a two-dimensional operator space with a nontrivial right M-summand.

We can say a little more about the multiplier algebras of two-dimensional spaces.

PROPOSITION 4.3. If X is a two-dimensional operator space, then $\mathcal{A}_\ell(X)$ is either $\mathbb{C}$, $\ell_2^\infty = \mathbb{C} \oplus \mathbb{C}$, or M_2.

(i) $\dim \mathcal{A}_\ell(X) = 4$ if and only if $X \cong C_2$ completely isometrically,

(ii) $\dim \mathcal{A}_\ell(X) = 2$ and $\dim \mathcal{A}_r(X) = 2$, if and only if $X \cong \ell_2^\infty$ completely isometrically,

(iii) X admits a nontrivial left M-projection P such that P^* is a right M-projection if and only if $X \cong C_2$ completely isometrically.

PROOF. The first assertion follows from Proposition 4.1.

(i) This follows from Theorem 5.54.

(ii) Suppose that P is a nontrivial left M-projection and that Q is a nontrivial complete right M-projection on X. Then $PQ = QP$ as maps on X. By elementary linear algebra it follows that $Q = P$ or $Q = \operatorname{Id} - P$. Thus P is both a left and a right M-projection, and hence is a complete (two-sided) M-projection. That is, $X \cong P(X) \oplus^\infty (\operatorname{Id} - P)(X) \cong \ell_2^\infty$ completely isometrically.

(iii) Suppose that P is a nontrivial left M-projection on X such that P^* is a right M-projection on X^*. Let $x, y \in X$ be unit vectors such that $Px = x$ and $Py = 0$. By Proposition 4.2, the map $T : C_2 \to X : \begin{bmatrix} a \\ b \end{bmatrix} \mapsto ax + by$ is a surjective complete contraction. Now let $\phi, \psi \in \operatorname{ball}(X^*)$ with $\phi(x) = \psi(y) = 1$.

Set $f = P^*(\phi)$ and $g = (\mathrm{Id} - P^*)(\psi)$. Then $P^*f = f$, $f(x) = 1$, and $f(y) = 0$. Likewise, $P^*g = 0$, $g(x) = 0$, and $g(y) = 1$. By the 'other-handed' version of Proposition 4.2, the map $S : R_2 \to X^* : \begin{bmatrix} \alpha & \beta \end{bmatrix} \mapsto \alpha f + \beta g$ is a surjective complete contraction. Since

$$\langle S^*(ax + by), \begin{bmatrix} \alpha & \beta \end{bmatrix} \rangle = \langle ax + by, \alpha f + \beta g \rangle = a\alpha + b\beta = \left\langle \begin{bmatrix} a \\ b \end{bmatrix}, \begin{bmatrix} \alpha & \beta \end{bmatrix} \right\rangle,$$

we have $S^* = T^{-1}$, and so T is a complete isometry. $\square$

4.2. MIN and MAX Spaces

We recall the two most common methods to regard a Banach space X as an operator space: namely $\mathrm{MIN}(X)$ and $\mathrm{MAX}(X)$. The following facts about a Banach space X may be found in [**BEZ**, Section 6.1]. Firstly, the left M-projections on $\mathrm{MIN}(X)$ are exactly the classical M-projections on X. Therefore, the right M-summands and right M-ideals of $\mathrm{MIN}(X)$ are just the M-summands and M-ideals of X, respectively. Likewise, the right L-projections on $\mathrm{MAX}(X)$ are the L-projections on X, which implies that the left L-summands of $\mathrm{MAX}(X)$ are the L-summands of X.

4.3. Hilbertian Operator Spaces

For a Hilbertian operator space H, every left M-projection on H is an orthogonal projection on H. Consequently, if J is a right M-summand of H, then its orthogonal complement and complementary right M-summand are one and the same.

Recall that an operator space X is *homogeneous* if $B(X) = CB(X)$ isometrically. This implies that every surjective isometry is a complete isometry, and for a Hilbertian operator space, the latter condition is equivalent to homogeneity (see [**Pis1**]).

THEOREM 4.4. *Let H be a homogeneous Hilbertian operator space, of dimension ≥ 2. Then the following are equivalent:*

   (i) *H is a Hilbert column space.*
   (ii) *H has a nontrivial right M-summand.*
   (iii) *H has a nontrivial left-adjointable multiplier (i.e. $\mathcal{A}_\ell(H) \neq \mathbb{C}$).*

PROOF. Clearly (i) $\Rightarrow$ (ii) $\Rightarrow$ (iii).

(iii) $\Rightarrow$ (ii) Since Banach space reflexivity implies operator space reflexivity, H is a dual operator space, and $\mathcal{A}_\ell(H)$ is a W^*-algebra by Theorem 2.3. Therefore, $\mathcal{A}_\ell(H)$ is the norm-closed linear span of its orthogonal projections (the left M-projections on H). Thus, if H has no nontrivial right M-summands, then $\mathcal{A}_\ell(H) = \mathbb{C}$.

(ii) $\Rightarrow$ (i) Suppose J is a nontrivial right M-summand of H, with corresponding left M-projection P. Let U be a unitary operator on H. Then U is a completely isometric surjection, and UPU^{-1} is a left M-projection on H, as discussed at the end of Section 2.2. Thus, every closed linear subspace of H of the same (Hilbertian) dimension as J is a right M-summand of H. Now let $x \in J$ and $y \in J^\perp$ be unit vectors. Define $K = J \cap \{x\}^\perp$. Then $\mathrm{span}\{y\} \vee K$ has the same dimension as $\mathrm{span}\{x\} \vee K = J$. Thus, $\mathrm{span}\{y\} \vee K$ is a right M-summand of H. By Proposition 5.23, $K = J \cap (\mathrm{span}\{y\} \vee K)$ is a right M-summand of H, which implies that

$K^\perp$ is as well. Again by Proposition 5.23, span$\{x\} = J \cap K^\perp$ is a right M-summand of H. Arguing as above, every one-dimensional linear subspace of H is a right M-summand of H. Hence, every rank-one orthogonal projection on H is an element of $\mathcal{A}_\ell(H)$. By polarization, every rank-one operator is an element of $\mathcal{A}_\ell(H)$. Therefore, if $\{e_i : i \in I\}$ is an orthonormal basis for H, then $\{e_i \otimes \bar{e}_i : i \in I\}$ is a family of mutually orthogonal and equivalent left M-projections which add to the identity in the point-norm topology. By Theorem 5.54, we have $H \cong C_I \cong H_c$. □

THEOREM 4.5. *If X is an operator space, then the following are equivalent:*

(i) *X is completely isometric to a Hilbert column space.*
(ii) *$B(X) = \mathcal{A}_\ell(X)$. That is, every bounded operator on X is a left adjointable multiplier.*
(iii) *$CB(X) = \mathcal{A}_\ell(X)$. That is, every completely bounded operator on X is a left adjointable multiplier.*
(iv) *$B(X) = \mathcal{M}_\ell(X)$ isometrically.*
(v) *$CB(X) = \mathcal{M}_\ell(X)$ isometrically.*
(vi) *Every one-dimensional subspace of X is a right M-summand.*

PROOF. (i) $\Rightarrow$ (ii) See e.g. [**B6**, Proposition 4.3 (ii)].

(ii) $\Rightarrow$ (iii) We have the inclusions $\mathcal{A}_\ell(X) \subset CB(X) \subset B(X) = \mathcal{A}_\ell(X)$.

(iii) $\Rightarrow$ (v) We have the isometric inclusion $CB(X) = \mathcal{A}_\ell(X) \subset \mathcal{M}_\ell(X)$ as well as the contractive inclusion $\mathcal{M}_\ell(X) \subset CB(X)$. It follows that $CB(X) = \mathcal{M}_\ell(X)$ isometrically.

(v) $\Rightarrow$ (vi) Let x be a unit vector in X, and $\psi \in X^*$ be such that $\|\psi\| = 1$ and $\psi(x) = 1$. Define $P : X \to X$ by $P(y) = \psi(y)x$ for all $y \in X$. Then P is a completely contractive projection. Since $CB(X) = \mathcal{M}_\ell(X)$ by assumption, P is a left multiplier of X and $\|P\|_{\mathcal{M}_\ell(X)} \leq 1$. It follows that P is a left M-projection on X, and that span$\{x\}$ is a right M-summand of X.

(vi) $\Rightarrow$ (i) Suppose that every one-dimensional subspace of X is a right M-summand of X. First suppose that $\dim(X) \geq 3$. Then every finite-dimensional subspace of X is a right M-summand of X, by Proposition 5.24. In particular, every two-dimensional subspace of X is contractively complemented. By a result of Kakutani and Bohnenblust ([**Kak, Boh**]), X is Hilbertian. Arguing as in the conclusion of the proof of Theorem 4.4, we see that $X = C_I$ completely isometrically, where I is any index set corresponding to an orthonormal basis for X. In the case that $\dim(X) = 2$ we first observe that (vi) implies that there exist non-commuting left M-projections on X. Hence $\dim(\mathcal{A}_\ell(X)) = 4$. Proposition 4.3 now gives $X = C_2$.

(ii) $\Rightarrow$ (iv) $\Rightarrow$ (vi) This is nearly identical to the proof of (iii) $\Rightarrow$ (v) $\Rightarrow$ (vi), and is left to the reader. □

Remark. We do not know whether the condition $B(X) = \mathcal{M}_\ell(X)$ (as sets) implies that X is Hilbert column space completely isometrically. If every bounded map on X is a left multiplier, then the open mapping theorem implies that $B(X)$ is isomorphic to the operator algebra $\mathcal{M}_\ell(X)$. It is known that this implies that X is isomorphic to a Hilbert space (see [**B1**, Theorem 3.4]). Also, it implies that X is λ-homogeneous for some λ. The latter conditions, while necessary, are not sufficient. For example, R_n (n-dimensional row space) is a homogeneous Hilbertian operator space, yet $\mathcal{M}_\ell(X) = \mathbb{C} \neq B(X)$.

4.4. C^*-Algebras

Let $\mathcal{A}$ be a C^*-algebra. As observed in [**BEZ**], §1, the left M-projections on $\mathcal{A}$ are precisely the maps

$$P : \mathcal{A} \to \mathcal{A} : T \mapsto ET,$$

where E is an orthogonal projection in $\mathcal{M}(\mathcal{A})$, the multiplier algebra of $\mathcal{A}$ (see [**Weg**], Chapter 2, for the definition of this algebra). More generally, $\mathcal{A}_\ell(\mathcal{A}) = \mathcal{M}(\mathcal{A})$ and $\mathcal{M}_\ell(\mathcal{A}) = \mathcal{LM}(\mathcal{A})$, the left multiplier algebra of $\mathcal{A}$ (cf. [**Weg**], Exercise 2.F). It follows that the right M-ideals of $\mathcal{A}$ are simply the closed right ideals. On the other hand, $\mathcal{A}$ has trivial left L-structure ([**BSZ**], Theorem 5.1). For further information on the one-sided M- and L-structure of C^*-algebras, see [**BSZ**]. See [**Eff1, Pro, Ake2, RN**] for some studies of the one-sided ideal structure of C^*-algebras, and the connection with important projections in the second dual.

4.5. Nonselfadjoint Operator Algebras

Let $\mathcal{B}$ be an operator algebra with a two-sided contractive approximate identity. In fact many of the results below only require a one-sided contractive approximate identity, but for simplicity we will stick to the two-sided case. It is shown in [**B6**] that $\mathcal{M}_\ell(\mathcal{B})$ is the usual operator algebra left multiplier algebra $\mathcal{LM}(\mathcal{B})$, and thus $\mathcal{A}_\ell(\mathcal{B}) = \{T \in \mathcal{LM}(\mathcal{B}) : T^* \in \mathcal{LM}(\mathcal{B})\}$. Consequently, the left M-projections on $\mathcal{B}$ are simply the maps of left multiplication by an orthogonal projection e in $\mathcal{LM}(\mathcal{B})$, the right M-summands of $\mathcal{B}$ are the principal right ideals $e\mathcal{B}$ for such e, and the right M-ideals of $\mathcal{B}$ are the closed right ideals having a contractive left approximate identity ([**BEZ**], Proposition 6.4).

Thus the one-sided M-ideals in a unital or 'approximately unital' operator algebra constitute an interesting class of one-sided ideals. It seems that there is very little that can be said in general about one-sided ideals in an operator algebra. Indeed, little is known about even the general ideals of commonly encountered function algebras. It is interesting, therefore, that the one-sided M-ideal theory suggests a tractable subclass for which some general theory is possible: namely the class of closed right ideals having a contractive left approximate identity. Such ideals arise quite naturally; we will demonstrate this at the end of this subsection by constructing some interesting examples. Furthermore, with some thought it becomes clear that the one-sided M-structure developed in the present article and in [**BEZ**] is key to the behavior of such right ideals in operator algebras, and of these approximate identities. That is, if one wishes to investigate how such right ideals behave with respect to basic constructions such as intersections or joins—the 'calculus' of such right ideals and their approximate identities, if you will—then this essentially boils down to the general calculus of one-sided M-ideals developed here. We will not systematically spell out all the applications to ideals in operator algebras of the theorems in the present article, we merely list some sample and typical applications. Some other relatively recent results concerning this class of right ideals in operator algebras, from the one-sided M-structure viewpoint, may be found in [**BSZ, B3, B5**] for example.

COROLLARY 4.6. Let $\mathcal{B}$ be an operator algebra with a two-sided contractive approximate identity. The closed span in $\mathcal{B}$ of a family of closed right ideals, each possessing a left contractive approximate identity for itself, is a right ideal also possessing a left contractive approximate identity.

PROOF. This follows from the fact that these are the right M-ideals in $\mathcal{B}$, together with Corollary 5.24. □

We are grateful to George Willis for some examples that show that the analogous result fails quite badly for Banach algebras. In fact it is not hard to find three- or four-dimensional counterexamples. Thus Corollary 4.6 is solely a 'one-sided M-ideal' phenomenon.

Results of the flavor of Corollary 4.6, but pertaining to 'intersections' rather than 'closed spans' may similarly be deduced immediately from the last five results in Section 5.4. We will not take the space to write these out explicitly here.

COROLLARY 4.7. Let $\mathcal{B}$ be an operator algebra with a two-sided contractive approximate identity, and fix an element a in the diagonal $\mathcal{B} \cap \mathcal{B}^*$ (or in the diagonal of $\mathcal{LM}(\mathcal{B})$). Let $J = a\mathcal{B} = \{ab : b \in \mathcal{B}\}$, a principal right ideal.

   (i) The norm closure of J is a right ideal possessing a left contractive approximate identity.
   (ii) If J is already closed, then $J = e\mathcal{B}$, for an orthogonal projection $e \in J$.

PROOF. Note that $J = \operatorname{ran}(T)$ where T is the map of left multiplication by a on B. This map is adjointable by remarks above.

   (i) Appeal to Corollary 3.15. Alternatively, this may be argued using ideals in the diagonal C^*-algebra.

   (ii) By Corollary 3.10 applied to the map $Tb = ab$, and the remarks above, we have $J = e\mathcal{B}$, for an orthogonal projection $e \in \mathcal{LM}(\mathcal{B})$. If $\mathcal{B}$ is unital then we are done, and note that in this case J has a left identity of norm one. If not we apply (ii) in the unital case but with $\mathcal{RM}(\mathcal{B})$ in place of $\mathcal{B}$. Note that $J \subset a\,\mathcal{RM}(\mathcal{B})$ clearly. For $T \in \mathcal{RM}(\mathcal{B})$, and (e_t) a contractive approximate identity of $\mathcal{B}$, we have $aT = \lim_t a e_t T \in J$. Thus $J = a\,\mathcal{RM}(\mathcal{B})$. We deduce from the unital case that J contains a left identity e of norm one. Thus $J = e\mathcal{B}$. □

We end this section with some examples showing that right ideals with a left contractive approximate identity come up quite naturally in various contexts. First, consider a unital operator algebra $\mathcal{A}$, and the subalgebra $J = C_\infty(R_\infty^w(\mathcal{A}))$ of $R_\infty^w(C_\infty(\mathcal{A}))$ (see Appendix B). It is easy to check that J is a right ideal with a left contractive approximate identity.

The second example below was suggested to us by G. A. Willis, and is related to a construction used in some of his papers (e.g. [**Wil**]). Let $\mathcal{A}$ be any unital operator algebra (or even, for the time being, a unital Banach algebra). Fix $a \in \mathcal{A}$, with $\|a\| = 1$. Let J be the closure of $(1-a)\mathcal{A}$. This is a right ideal, and J will have a left contractive approximate identity if the norms of the elements $e_n = 1 - \frac{1}{n}\sum_{k=1}^{n} a^k$ approach 1 (or have a subnet whose norms approach 1). By a simple algebraic computation it is easy to check that $e_n \in (1-a)\mathcal{A}$, and that $e_n(1-a) \to (1-a)$. In fact it is not hard to ensure that $\|e_n\| \to 1$. For example, this will hold by basic spectral theory if $\mathcal{A}$ is an operator algebra, and if a is normal, with spectrum excluding a deleted neighborhood of 1.

4.6. Hilbert C^*-Modules

Let X be a right Hilbert C^*-module (see e.g. [**Lan**]). Then the right M-ideals of X are the closed right submodules of X, and the right M-summands of X are the orthogonally complemented right submodules of X. For the details, see [**BEZ**], where we also give one or two applications of one-sided M-ideal theory to C^*-modules. We will not explicitly list more such applications in the present memoir. In fact, the flow is mostly in the other direction: most of the results in the present memoir may be seen as *generalizations* of known results about C^*-modules. We invite the reader to glance once through our results with this in mind; the parallels are quite striking. This perspective is highlighted in [**B4**].

4.7. Operator Modules

We will not define operator modules here, but simply give the equivalent characterization discovered in [**B6**] (see [**BP**] for an alternative account): they correspond to the module actions given by the formula $ax = \pi(a)(x)$, for $a \in \mathcal{A}, x \in X$, for a completely contractive unital homomorphism $\pi : \mathcal{A} \to \mathcal{M}_\ell(X)$. Here $\mathcal{A}$ is an operator space and a unital algebra. If $\mathcal{A}$ is a C^*-algebra we may replace $\mathcal{M}_\ell(X)$ by $\mathcal{A}_\ell(X)$ here. Thus for example from a result in [**BSZ**] one sees that there are no nontrivial (i.e. nonscalar) operator module actions of a C^*-algebra on an L^1 space, or more generally on the predual of a von Neumann algebra. This is because for such spaces all left adjointable maps are trivial (i.e. a scalar multiple of the identity). On the other hand such 'L^1 spaces' X have rich one-sided Cunningham algebras, and thus X is a nontrivial bimodule over $\mathcal{C}_\ell(X)$ and $\mathcal{C}_{\mathrm{r}}(X)$.

PROPOSITION 4.8. *If J is a left M-ideal in an operator space X, then $T(J) \subset J$ for all $T \in \mathcal{M}_\ell(X)$.*

PROOF. Let P be the right M-projection onto $J^{\perp\perp}$. Inside X^{**} we have

$$T(J) = T^{**}(J) \subset T^{**}(P(X^{**})) = P(T^{**}(X^{**})) \subset J^{\perp\perp},$$

using the fact from Section 5.3 that $T^{**} \in \mathcal{M}_\ell(X^{**})$, and the last point in Section 2.1. Thus $T(J) \subset J^{\perp\perp} \cap X = J$, by basic functional analysis. $\square$

COROLLARY 4.9. *Let X be a left operator $\mathcal{A}$-module over an algebra $\mathcal{A}$ as above. If J is a left M-ideal in X, then J is an $\mathcal{A}$-submodule of X.*

From this result and the situation in Section 4.6, one might expect that every submodule of a left operator module X is a left M-ideal. However this is not true in general. In well-behaved situations like when X is an operator algebra with contractive approximate identity one at least needs an extra condition (see Section 4.5). Another good example of this is furnished by the strong Morita equivalence $\mathcal{A}$-$\mathcal{B}$-bimodules of [**BMP**]. These are the analogues for nonselfadjoint operator algebras of the C^*-algebraic strong Morita equivalence bimodules of Rieffel (see [**Lan**] for example), or equivalently the 'full' Hilbert C^*-modules discussed in Section 4.6. By analogy with the C^*-module case one would expect that if X is a strong Morita equivalence $\mathcal{A}$-$\mathcal{B}$-bimodule in the sense of [**BMP**], then the right M-ideals of X are exactly the closed right $\mathcal{B}$-submodules. However in fact one needs to impose further conditions, and even then the situation is quite complicated—see the discussion in [**B4**, Section 10]. The right M-summands of such a bimodule are easier to deal with, they are just the ranges of completely contractive idempotent maps on X which are right $\mathcal{B}$-module maps. That is, they are exactly the analogues in this situation of the 'orthogonally complemented' submodules. The proof of this right M-summand case (from joint work of the first author with Solel from '00) may be found in [**B4**, Section 10]; it proceeds by first establishing that the 'Hamana triple envelope' or 'noncommutative Shilov boundary' of X is a canonical C^*-algebraic strong Morita equivalence bimodule over the C^*-envelopes of $\mathcal{A}$ and $\mathcal{B}$.

4.8. Operator Systems and M-Projective Units

If X is an operator system, then the left M-projections are just the 'involutions' of the right M-projections, where the 'involution' of a map T on X is the map $x \mapsto T(x^*)^*$. Indeed, it is clear from the definition of the multiplier algebras in terms of $C_e^*(X)$ (see e.g. [**B6**], or [**B3**, p. 505–506]) that $\mathcal{A}_\ell(X)$ and $\mathcal{A}_r(X)$ may be viewed as *subsets* of X, and in fact these two subsets of X coincide. We are not saying that every left adjointable multiplier is right adjointable, this would imply that every closed left ideal of a unital C^*-algebra is a closed right ideal! In fact, the left M-structure of an operator system is related to the right M-structure by the above-mentioned involution. Since this will not play a role for us, we leave the details to the reader.

We remark in passing that for a certain class of operator systems, multipliers were investigated by Kirchberg [**Kir**]. This is related to his study of the sum $J + K$ of a left and a right ideal in a C^*-algebra, which was important in his profound work on nuclear C^*-algebras, etc. For general operator systems, and in fact for a large class of ordered operator spaces, one-sided multipliers were introduced by W. Werner. Later it was noticed that there is a way to view one-sided multipliers on arbitrary operator spaces within this context (see [**WWer**]).

We say that an element x in an operator system X is a *left M-projective unit*, if $x = Pe$ for a left M-projection P (here e is the unit in X). We write X_1^+ for the positive part of the unit ball of X.

COROLLARY 4.10. *Let X be an operator system which is also a dual operator space. Then X has a product with respect to which it is a W^*-algebra (with the same underlying operator system structure), if and only if every extreme point of X_1^+ is a left M-projective unit in X. In this case, such a product is unique.*

PROOF. The uniqueness of the product may be deduced from any operator space variant of the Banach-Stone theorem (see e.g. [**ER4**, Corollary 5.2.3]).

($\Rightarrow$) If X is a W^*-algebra, then its left M-projective units are exactly its orthogonal projections, by facts in Section 4.4. Thus the result follows from the characterization of extreme points in the positive part of the unit ball of a W^*-algebra [**Sak2**, Proposition 1.6.2].

($\Leftarrow$) Consider the map $\rho : \mathcal{A}_\ell(X) \to X : T \mapsto Te$. It is explained above Theorem 1.9 in [**B3**] that ρ is a complete isometry, and it is easy to see that ρ is weak-* continuous. Hence, by the Krein-Smulian Theorem A.1, we see that ρ is a weak-* homeomorphism with weak-* closed range. By hypothesis, ran(ρ) contains all extreme points of X_1^+. By the Krein-Milman Theorem, ran(ρ) contains X_1^+, and hence ρ maps onto X. $\square$

This result was inspired by, and is closely related to, an old theorem of K. H. Werner [**KWer**, Corollary 5.1].

4.9. Locally Reflexive Operator Spaces

For results on the one-sided M-structure of locally reflexive operator spaces, we refer the reader to a paper in preparation by the second author [**Z2**]. The main result is that one-sided M-ideals in such spaces can be characterized in terms of 'local' one-sided M-projections. This generalizes classical results of Lima [**Lim**].

CHAPTER 5

Constructions

In this chapter we develop the heart of the calculus of one-sided M-ideals. It is quite interesting that there are perfect analogs of approximately half the results in the classical M-ideal calculus. For the other half, some additional and natural hypotheses need to be imposed.

It would take up too much space to thoroughly define and introduce every one of the constructions listed in this chapter. Readers interested in a particular construction should consult the basic texts on operator spaces, or the original papers, for precise definitions and basic facts.

5.1. Opposite and Conjugate

If X is an operator space, then X^{op} is defined to be the same Banach space, but with 'transposed' matrix norms: $\|[x_{ij}]\|_{\mathrm{op}} \equiv \|[x_{ji}]\|$. Then X^{op} is an operator space, as is well known and easy to see. If $\sigma : X \to \mathcal{A}$ is a complete isometry into a C^*-algebra, then we obtain a matching complete isometry $\sigma^{\mathrm{op}} : X^{\mathrm{op}} \to \mathcal{A}^{\mathrm{op}}$. The latter space is a C^*-algebra, namely $\mathcal{A}$ with its reversed multiplication. Similarly, if $\mathcal{B}$ is an operator algebra, a subalgebra of a C^*-algebra $\mathcal{A}$, then with the reversed product, $\mathcal{B}^{\mathrm{op}}$ is completely isometrically homomorphic to the obvious subalgebra of the C^*-algebra $\mathcal{A}^{\mathrm{op}}$. From this it is easy but tedious to check that $\mathcal{M}_\ell(X^{\mathrm{op}}) \cong \mathcal{M}_{\mathrm{r}}(X)^{\mathrm{op}}$ and $\mathcal{A}_\ell(X^{\mathrm{op}}) \cong \mathcal{A}_{\mathrm{r}}(X)^{\mathrm{op}}$ canonically. We sketch the proof, leaving many details to the reader, since we shall not need it. It is easy to check that the canonical map is an isometric homomorphism from $\mathcal{M}_\ell(X^{\mathrm{op}})$ to $\mathcal{M}_{\mathrm{r}}(X)^{\mathrm{op}}$. The complete isometry hinges on the fact that $M_n(X^{\mathrm{op}}) \cong M_n(X)^{\mathrm{op}}$ completely isometrically, via 'transposition', and that $M_n(\mathcal{M}_\ell(X)) \cong \mathcal{M}_\ell(M_n(X))$ canonically. Using these two facts, we see that $M_n(\mathcal{M}_\ell(X^{\mathrm{op}})) \cong \mathcal{M}_\ell(M_n(X)^{\mathrm{op}})$. By the isometric case, the latter algebra is isometrically homomorphic to

$$\mathcal{M}_{\mathrm{r}}(M_n(X))^{\mathrm{op}} \cong M_n(\mathcal{M}_{\mathrm{r}}(X))^{\mathrm{op}} \cong M_n(\mathcal{M}_{\mathrm{r}}(X)^{\mathrm{op}}).$$

Thus $M_n(\mathcal{M}_\ell(X^{\mathrm{op}})) \cong M_n(\mathcal{M}_{\mathrm{r}}(X)^{\mathrm{op}})$, which establishes the complete isometry. The $\mathcal{A}_\ell(\cdot)$ case is easily derived from the $\mathcal{M}_\ell(\cdot)$ case.

It is thus easy to see that if J is a right M-ideal or right M-summand of X, then J is a left M-ideal or left M-summand of X^{op}. Likewise, left L-summands of X correspond to right L-summands of X^{op}.

Similar assertions hold for the *conjugate operator space* $\overline{X}$, which is defined to be the set $\{\overline{x} : x \in X\}$ together with the linear structure defined by

$$\overline{x} + \overline{y} \equiv \overline{x + y} \text{ and } \alpha \overline{x} \equiv \overline{\overline{\alpha}x},$$

and the matrix norms

$$\|[\overline{x}_{ij}]\| \equiv \|[x_{ij}]\|.$$

We have, for example, $\mathcal{M}_\ell(\overline{X}) \cong \overline{\mathcal{M}_\ell(X)}$ completely isometrically and homomorphically, and $\mathcal{A}_\ell(\overline{X}) \cong \overline{\mathcal{A}_\ell(X)}$ *-homomorphically. Thus, if J is a right M-ideal or right M-summand of X, then $\overline{J} \equiv \{\overline{x} : x \in J\}$ is a right M-ideal or right M-summand of $\overline{X}$. Likewise, the left L-summands of X correspond (via 'conjugation') to the left L-summands of $\overline{X}$.

From the facts above for the 'opposite' and 'conjugate', it is easy to determine the one-sided multipliers and M-ideals of the *adjoint operator space* $X^\star$, which is defined to be $(\overline{X})^{\mathrm{op}}$ (see e.g. [**BLM**]).

We say that an operator space is *symmetric* if $X = X^{\mathrm{op}}$. For a symmetric operator space it is easy to see that $T \in \mathcal{M}_\ell(X)$ if and only if $T \in \mathcal{M}_\mathrm{r}(X)$, and similarly for $\mathcal{A}_\ell(X)$ and $\mathcal{A}_\mathrm{r}(X)$. It follows from the last fact in Section 2.1 that $\mathcal{A}_\ell(X)$ and $\mathcal{A}_\mathrm{r}(X)$ are commutative. In fact it follows from Section 7.1 that $\mathcal{A}_\ell(X) = \mathcal{A}_\mathrm{r}(X) = Z(X)$, with the notation from that Section. Thus from that Section we see that the left M-projections (resp. right M-summands, right M-ideals) on X are precisely the 'complete M-projections' (resp. complete M-summands, complete M-ideals).

Examples of symmetric operator spaces include the $O\ell^p$ spaces of Pisier [**Pis1**]. These are formed by interpolating between the symmetric spaces $O\ell^1 = c_0^*$ and ℓ^∞. Thus they are symmetric. In this case there are no nontrivial classical M-ideals or M-projections if $p < \infty$, hence no nontrivial complete M-ideals or complete M-projections, and hence no nontrivial one-sided M-ideals or one-sided M-projections. Since in this case $\mathcal{A}_\ell(O\ell^p)$ is a W^*-algebra and therefore generated by its projections, we must have that $\mathcal{A}_\ell(O\ell^p)$ is one-dimensional.

5.2. Subspace and Quotient

To facilitate our development, we introduce two operators associated with an operator and an invariant subspace:

DEFINITION 5.1. Let X be an operator space, Y be a closed linear subspace of X, and $T : X \to X$ be a linear map such that $T(Y) \subset Y$. By $T|_Y : Y \to Y$ we denote the restriction of T to Y, and by $T/Y : X/Y \to X/Y$ we denote the (well-defined) linear map

$$(T/Y)(x + Y) = Tx + Y.$$

We observe that if T is completely bounded, then so are $T|_Y$ and T/Y, and $\|T|_Y\|_{cb}, \|T/Y\|_{cb} \leq \|T\|_{cb}$. The latter inequality is a consequence of the calculation

$$\begin{aligned}
\|Tx + Y\| &= \inf\{\|Tx + y\| : y \in Y\} \leq \inf\{\|T(x+y)\| : y \in Y\} \\
&\leq \|T\| \inf\{\|x + y\| : y \in Y\} = \|T\|\|x + Y\|,
\end{aligned}$$

as well as the fact that under the isometric isomorphism $M_n(X/Y) = M_n(X)/M_n(Y)$, $(T/Y)_n$ is identified with $T_n/M_n(Y)$.

PROPOSITION 5.2 ([**BEZ**], Proposition 5.8). Let X be an operator space, Y be a closed linear subspace of X, and $T : X \to X$ be a linear map such that $T(Y) \subset Y$.

(i) If $T \in \mathcal{M}_\ell(X)$, with $\|T\|_{\mathcal{M}_\ell(X)} \leq 1$, then $T|_Y \in \mathcal{M}_\ell(Y)$ and $T/Y \in \mathcal{M}_\ell(X/Y)$, with $\|T|_Y\|_{\mathcal{M}_\ell(Y)} \leq 1$ and $\|T/Y\|_{\mathcal{M}_\ell(X/Y)} \leq 1$.

(ii) If $T \in \mathcal{A}_\ell(X)$ and $T^\star(Y) \subset Y$, then $T|_Y \in \mathcal{A}_\ell(Y)$ and $T/Y \in \mathcal{A}_\ell(X/Y)$. Furthermore, $(T|_Y)^\star = T^\star|_Y$ and $(T/Y)^\star = T^\star/Y$.

We begin by examining subspaces. Our first result is a hereditary property of right M-summands (resp. right M-ideals):

PROPOSITION 5.3. *Let X be an operator space, J be a closed linear subspace of X, and Y be a closed linear subspace of J.*

(i) *If Y is a right M-summand of X, then Y is a right M-summand of J.*

(ii) *If Y is a right M-ideal of X, then Y is a right M-ideal of J.*

PROOF. (i) Let $P \in \mathcal{A}_\ell(X)_{sa}$ be the left M-projection onto Y. Then $P(J) = Y \subset J$. Therefore, by Proposition 5.2, $P|_J \in \mathcal{A}_\ell(J)_{sa}$. Clearly, $P|_J$ is an idempotent with range Y. Thus, Y is a right M-summand of J.

(ii) By (i), $Y^{\perp\perp}$ is a right M-summand of $J^{\perp\perp}$. It follows that $Y^{\perp\perp}$ is a right M-summand of J^{**}, which says that Y is a right M-ideal of J. □

On the other hand, if Y is a right M-ideal of J and J is a right M-ideal of X, then it need not be the case that Y is a right M-ideal of X (compare with [**HWW**], Proposition I.1.17). The following example of this phenomenon is due to K. Kirkpatrick (an R.E.U. student advised by the first author):

EXAMPLE 5.4. Let

$$X = \left\{ \begin{bmatrix} a & 0 \\ b & 0 \\ 0 & \frac{a+b}{2} \\ 0 & c \end{bmatrix} : a, b, c \in \mathbb{C} \right\} \subset M_{4,2}.$$

Define

$$J = \left\{ \begin{bmatrix} a & 0 \\ b & 0 \\ 0 & \frac{a+b}{2} \\ 0 & 0 \end{bmatrix} : a, b \in \mathbb{C} \right\} \quad \text{and} \quad Y = \left\{ \begin{bmatrix} a & 0 \\ 0 & 0 \\ 0 & \frac{a}{2} \\ 0 & 0 \end{bmatrix} : a \in \mathbb{C} \right\}.$$

It is easy to see that J is a right M-summand of X, and that the 'natural' projection P of X onto J is the corresponding left M-projection. Likewise, Y is a right M-summand of J, and the natural projection Q of J onto Y is the corresponding left M-projection. This follows from the fact that the map

$$C_2 \to J : \begin{bmatrix} a \\ b \end{bmatrix} \mapsto \begin{bmatrix} a & 0 \\ b & 0 \\ 0 & \frac{a+b}{2} \\ 0 & 0 \end{bmatrix}$$

is a completely isometric isomorphism. For future reference, let $\tilde{J}$ be the complementary right M-summand of J in X (namely, $\tilde{J}$ is the copy of $\mathbb{C}$ in the $4 - 2$ corner of X), and let $\tilde{Y}$ be the complementary right M-summand of Y in J. Now suppose for the purpose of a contradiction that Y is a right M-summand of X. Since $E = Q \circ P : X \to X$ is a contractive projection with range Y, it must be the left M-projection of X onto Y (cf. the fourth 'bullet' in Section 2.3). But $\nu_E^c : X \to C_2(X)$ is not isometric, as can be seen by applying it to

$$x_\lambda = \begin{bmatrix} 1 & 0 \\ 1 & 0 \\ 0 & 1 \\ 0 & \lambda \end{bmatrix}$$

for λ sufficiently large. Indeed, $\|\nu_E^c(x_\lambda)\| = \max\left\{\sqrt{2}, \sqrt{\frac{1}{2} + \lambda^2}\right\}$ whereas $\|x_\lambda\| = \max\{\sqrt{2}, \sqrt{1 + \lambda^2}\}$.

We remark that by symmetry, $\tilde{Y}$ is also a right M-summand of J which fails to be a right M-summand of X.

Before moving on to quotients, we identify the 'culprit' behind Example 5.4. Namely, we prove in Proposition 5.6 below that if J is a right M-summand of a dual operator space X, then all right M-summands of J are right M-summands of X if and only if $\mathcal{A}_\ell(J) \cong P\mathcal{A}_\ell(X)P$, where P is the left M-projection of X onto J. In the case of Example 5.4, $\mathcal{A}_\ell(J) = M_2$ whereas $P\mathcal{A}_\ell(X)P = \ell_2^\infty$. We begin with an important lemma.

LEMMA 5.5. *Let X be an operator space, let P be a left M-projection on X, and let ρ be the map $T \mapsto T|_{P(X)}$ from $P\mathcal{A}_\ell(X)P$ into $\mathcal{A}_\ell(P(X))$.*

    (i) *ρ identifies the C^*-algebra $P\mathcal{A}_\ell(X)P$ $*$-isomorphically with a (possibly proper) C^*-subalgebra of $\mathcal{A}_\ell(P(X))$.*

    (ii) *If X is a dual operator space, then ρ is a $*$-isomorphism from $P\mathcal{A}_\ell(X)P$ onto a W^*-subalgebra of $\mathcal{A}_\ell(P(X))$, and ρ is a weak-$*$ homeomorphism.*

PROOF. Let ρ be the above map, which is clearly a one-to-one homomorphism into $B(P(X))$. By Proposition 5.2 we see that ρ does indeed map into $\mathcal{A}_\ell(P(X))$, and is a $*$-homomorphism. Hence ρ is also isometric and has closed range. The remark before the Lemma shows that this range may be a proper C^*-subalgebra. This gives (i).

If X is a dual operator space then $P(X)$ is weak-$*$ closed, by a remark after Theorem 2.5. Then (ii) follows from (i), the Krein-Smulian Theorem A.1, and Theorem 2.3.   □

PROPOSITION 5.6. *Let X be a dual operator space with a right M-summand J and corresponding left M-projection P. Then every right M-summand of J is a right M-summand of X if and only if $\mathcal{A}_\ell(J) \cong P\mathcal{A}_\ell(X)P$ via the map ρ in Lemma 5.5.*

PROOF. ($\Leftarrow$) This implication does not need X to be a dual space. Suppose that Q is a left M-projection on J. By assumption, $Q = \rho(T)$ for some $T \in P\mathcal{A}_\ell(X)P \subset \mathcal{A}_\ell(X)$. Since ρ is an injective *-homomorphism, T is a left M-projection on X. We have $T(X) = TP(X) = QP(X) = \mathrm{ran}(Q)$. Hence $\mathrm{ran}(Q)$ is a right M-summand of X.

($\Rightarrow$) Because the orthogonal projections generate any von Neumann algebra, we need only prove that every left M-projection Q on J equals $PRP|_J$ for a left M-projection R on X. By hypothesis, there exists a left M-projection R on X such that $\mathrm{ran}(R) = \mathrm{ran}(Q)$. Evidently, $PRP|_J$ is a left M-projection on J with range $\mathrm{ran}(Q)$. It follows from the uniqueness of left M-projections with a given range that $Q = PRP|_J$.   □

Perhaps the most frustrating problem in the theory of one-sided M-structure is that the equivalent conditions in the last Proposition may fail (see Example 5.4). In any case, this 'problem' suggests a definition:

DEFINITION 5.7. *A right M-summand J of an operator space X is called hereditary if every right M-summand of J is also a right M-summand of X.*

One might then ask for a reasonable sufficient condition for a right M-summand to be hereditary, and a first guess might be that all M-projections involved commute. However a simple modification of Example 5.4, with the 'a-b column' replaced by $diag\{a,b\}$, shows that such 'commutativity' does not help. (We remark that for this modified example it is quickest to compute the multiplier algebras explicitly from their formulation in terms of the 'triple envelope' (see [**Ham, B6**]); in this case the triple envelope is easily seen to be $\ell_2^\infty \oplus C_2$.)

The following is a rather strong sufficient condition. We say that a closed subspace J of an operator space X is a *Shilov subspace* if there is a triple envelope Z of X such that the triple subsystem of Z generated by J (that is, the smallest closed subspace of Z containing J and closed under the 'triple product') is a triple envelope of J. By the universal property of the triple envelope (see e.g. [**Ham**]), one can show that this notion does not depend on the particular triple envelope Z above.

PROPOSITION 5.8. *If a right M-summand J of an operator space X is a Shilov subspace of X, then J is hereditary.*

PROOF. Let P be the left M-projection of X onto J. If J is a Shilov subspace, then the restriction of the 'Shilov inner product' on X (see Section 3.4) to J is a 'Shilov inner product' for J. If Q is a left M-projection on J then Q is a projection in $\mathcal{A}_\ell(J)$, and so by (3.1) we have

$$\langle Qx, y \rangle = \langle x, Qy \rangle, \qquad x, y \in J.$$

Since $PQ = Q$, we have for $x, y \in X$ that

$$\langle QPx, y \rangle = \langle QPx, Py \rangle = \langle Px, QPy \rangle = \langle x, QPy \rangle.$$

Thus QP is a left adjointable contractive projection on X with range equal to the range of Q, which establishes the asserted statement. $\qquad\qquad\square$

Turning to quotients, we have the following positive result, which tells us that right M-summands and right M-ideals are stable under the taking of quotients.

PROPOSITION 5.9. *Let X be an operator space, J be a closed linear subspace of X, and Y be a closed linear subspace of J.*

   (i) *If J is a right M-summand of X, then J/Y is a right M-summand of X/Y.*
   (ii) *If J is a right M-ideal of X, then J/Y is a right M-ideal of X/Y.*

PROOF. (i) Let $P \in \mathcal{A}_\ell(X)_{sa}$ be the left M-projection onto J. Then $P(Y) = Y$. Therefore, by Proposition 5.9, $P/Y \in \mathcal{A}_\ell(X/Y)_{sa}$. Clearly, P/Y is an idempotent with range J/Y. Thus, J/Y is a right M-summand of X/Y.

   (ii) By (i), $J^{\perp\perp}/Y^{\perp\perp}$ is a right M-summand of $X^{**}/Y^{\perp\perp}$. It follows that $(J/Y)^{\perp\perp}$ is a right M-summand of $(X/Y)^{**}$, which says that J/Y is a right M-ideal of X/Y. $\qquad\qquad\square$

On the other hand, not every right M-ideal of a quotient of X arises as the image of a right M-ideal of X under the quotient map (compare again with [**HWW**], Proposition I.1.17). An instance of this phenomenon is furnished by the following example:

EXAMPLE 5.10. Let X, J, $\tilde{J}$, Y, and $\tilde{Y}$ have the same meanings as in Example 5.4. We will show that $(Y + \tilde{J})/\tilde{J}$ is a right M-summand of $X/\tilde{J}$, but that there does not exist a right M-summand K of X such that $(Y + \tilde{J})/\tilde{J} = K/\tilde{J}$. To prove the first assertion, note that under the canonical completely isometric isomorphism $X/\tilde{J} = J$, $(Y + \tilde{J})/\tilde{J}$ is identified with Y. Since Y is a right M-summand of J, $(Y+\tilde{J})/\tilde{J}$ is a right M-summand of $X/\tilde{J}$. To prove the second assertion, we proceed by contradiction. So let K be a right M-summand of X such that $K/\tilde{J} = (Y+\tilde{J})/\tilde{J}$. Replacing K by $K + \tilde{J}$, which is still a right M-summand of X by Proposition 3.3 for example, we may suppose that K contains $\tilde{J}$. It is easy to see that K must equal $Y + \tilde{J}$. Let F be the left M-projection of X onto $Y + \tilde{J}$. Since $\tilde{J} \subset Y + \tilde{J}$, we have $P^\perp F = FP^\perp = P^\perp$. Therefore, if

$$F\left(\begin{bmatrix} 0 & 0 \\ 1 & 0 \\ 0 & \frac{1}{2} \\ 0 & 0 \end{bmatrix}\right) = \begin{bmatrix} a & 0 \\ 0 & 0 \\ 0 & \frac{a}{2} \\ 0 & c \end{bmatrix},$$

then we may conclude that $c = 0$. On the other hand, the fact that $\nu_F^c : X \to C_2(X)$ is (completely) isometric allows us to conclude that $a = 0$. That is, F is the 'natural' projection of X onto $Y + \tilde{J}$. But then $F^\perp$ is the natural projection of X onto $\tilde{Y}$, and so $\tilde{Y}$ is a right M-summand of X. This contradicts the remark following Example 5.4.

The behavior of one-sided L-structure with respect to subspaces and quotients is similar. This is a consequence of the following lemma.

LEMMA 5.11. *Let X be an operator space, Y be a closed linear subspace of X, and $T : X \to X$ be a linear map such that $T(Y) \subset Y$. Then*

(i) *Under the completely isometric isomorphism $Y^* = X^*/Y^\perp$, $(T|_Y)^* : Y^* \to Y^*$ is identified with $T^*/Y^\perp : X^*/Y^\perp \to X^*/Y^\perp$.*

(ii) *Under the completely isometric isomorphism $(X/Y)^* = Y^\perp$, $(T/Y)^* : (X/Y)^* \to (X/Y)^*$ is identified with $T^*|_{Y^\perp} : Y^\perp \to Y^\perp$.*

The proof of this lemma is an easy exercise which we omit.

COROLLARY 5.12. *Let X be an operator space, Y be a closed linear subspace, and $T \in \mathcal{C}_r(X)$. If $T(Y) \subset Y$ and $T^\star(Y) \subset Y$, then $T|_Y \in \mathcal{C}_r(Y)$ and $T/Y \in \mathcal{C}_r(X/Y)$. Furthermore, $(T|_Y)^\star = T^\star|_Y$ and $(T/Y)^\star = T^\star/Y$.*

PROOF. Recall from Section 2.4 that for any operator space V, the isometric homomorphism $CB(V)^{\mathrm{op}} \to CB(V^*) : T \mapsto T^*$ restricts to a *-isomorphism $\mathcal{C}_r(V) \to \mathcal{A}_\ell(V^*)$ which is also a weak-* homeomorphism. Therefore, to show that $T|_Y \in \mathcal{C}_r(Y)$, it suffices to show that $(T|_Y)^* \in \mathcal{A}_\ell(Y^*)$. Likewise, to show that $T/Y \in \mathcal{C}_r(X/Y)$, it suffices to show that $(T/Y)^* \in \mathcal{A}_\ell((X/Y)^*)$. By Lemma 5.11, it is equivalent to show that $T^*/Y^\perp \in \mathcal{A}_\ell(X^*/Y^\perp)$ and $T^*|_{Y^\perp} \in \mathcal{A}_\ell(Y^\perp)$, respectively. But this follows from Proposition 5.2, as do the formulas appearing in the statement of the corollary. □

COROLLARY 5.13. *Let X be an operator space, J be a closed linear subspace of X, and Y be a closed linear subspace of J.*

(i) *If Y is a left L-summand of X, then Y is a left L-summand of J.*

(ii) *If J is a left L-summand of X, then J/Y is a left L-summand of X/Y.*

To prove this corollary one mimics the proofs of Propositions 5.3 and 5.9, with the role of Proposition 5.2 being played by Corollary 5.12. We leave the details to the interested reader.

5.3. Dual and Bidual

In this section we will establish some fundamental facts about one-sided multipliers and M-ideals of dual and bidual operator spaces. Our first goal is to show that the linear isometry $\Phi : CB(X) \to CB(X^{**}) : T \mapsto T^{**}$ restricts to a $*$-homomorphism $\mathcal{A}_\ell(X) \to \mathcal{A}_\ell(X^{**})$.

PROPOSITION 5.14. Let $T : X \to X$ be a linear map on an operator space. Then T is an element of the closed unit ball of $\mathcal{M}_\ell(X)$ if and only if T^{**} is an element of the closed unit ball of $\mathcal{M}_\ell(X^{**})$.

PROOF. By two applications of Theorem 2.1 we have that $\|T\|_{\mathcal{M}_\ell(X)} \leq 1$ if and only if

$$\|\tau_T^c\|_{cb} \leq 1 \Leftrightarrow \|(\tau_T^c)^{**}\|_{cb} \leq 1 \Leftrightarrow \|\tau_{T^{**}}^c\|_{cb} \leq 1 \Leftrightarrow \|T^{**}\|_{\mathcal{M}_\ell(X^{**})} \leq 1.$$

(A direct proof without using Theorem 2.1 may also be given). □

Thus the map Φ above restricts to a unital isometric homomorphism $\mathcal{M}_\ell(X) \to \mathcal{M}_\ell(X^{**})$. It is easy to check that a unital contraction between unital Banach algebras takes Hermitian elements to Hermitian elements. For a unital subalgebra $\mathcal{B}$ of a unital C^*-algebra $\mathcal{A}$ it is clear that the Hermitian elements of $\mathcal{B}$ are the selfadjoint elements of $\mathcal{A}$ which are in $\mathcal{B}$, or equivalently the selfadjoint elements of the diagonal of $\mathcal{B}$. Since $\mathcal{A}_\ell(X)$ is the diagonal of $\mathcal{M}_\ell(X)$ (cf. Section 2.3), it follows from the last few facts that Φ takes selfadjoint elements in $\mathcal{A}_\ell(X)$ to selfadjoint elements in $\mathcal{A}_\ell(X^{**})$. Hence Φ restricts to an isometric $*$-homomorphism from $\mathcal{A}_\ell(X)$ into $\mathcal{A}_\ell(X^{**})$. Consequently, $\mathcal{A}_\ell(X)$ may be regarded as a unital C^*-subalgebra of $\mathcal{A}_\ell(X^{**})$. Typically, the inclusion $\mathcal{A}_\ell(X) \subset \mathcal{A}_\ell(X^{**})$ is proper.

A simple consequence of the preceding discussion is the following result, which we will make use of in a later remark.

COROLLARY 5.15. Let X be an operator space, and let $T \in \mathcal{A}_\ell(X)$. Then $|T|^{**} = |T^{**}|$.

PROOF. We have observed that the map $\mathcal{A}_\ell(X) \to \mathcal{A}_\ell(X^{**}) : T \mapsto T^{**}$ is a $*$-homomorphism. Thus $|T|^{**}$ is positive in the C^*-algebra $\mathcal{A}_\ell(X^{**})$. We also deduce that

$$(|T|^{**})^2 = (T^\star T)^{**} = (T^{**})^\star T^{**}.$$

Hence $|T|^{**} = |T^{**}|$. □

We have seen that that $T \in \mathcal{M}_\ell(X)$ if and only if $T^{**} \in \mathcal{M}_\ell(X^{**})$. Likewise, we saw earlier that if $T \in \mathcal{A}_\ell(X)$, then $T^{**} \in \mathcal{A}_\ell(X^{**})$. It is natural to expect therefore that if $T^{**} \in \mathcal{A}_\ell(X^{**})$, then $T \in \mathcal{A}_\ell(X)$. In fact this is not true in general. Indeed, if $\mathcal{A}$ is a C^*-algebra and $A \in \mathcal{LM}(\mathcal{A}) \backslash \mathcal{M}(\mathcal{A})$, then the map $L_A : \mathcal{A} \to \mathcal{A}$ of left multiplication by A is a left multiplier of $\mathcal{A}$ which is not left adjointable. On the other hand, $L_A^{**} : \mathcal{A}^{**} \to \mathcal{A}^{**}$ is again the map of left multiplication by A, which is a left adjointable multiplier of $\mathcal{A}^{**}$ (since $A \in \mathcal{A}^{**}$). The following proposition clarifies the situation.

PROPOSITION 5.16. Let X be an operator space and $T : X \to X$ be bounded and linear. Then

(i) $T \in \mathcal{A}_\ell(X)_{sa}$ if and only if $T^{**} \in \mathcal{A}_\ell(X^{**})_{sa}$.
(ii) $T \in \mathcal{A}_\ell(X)$ if and only if $T^{**} \in \mathcal{A}_\ell(X^{**})$ and $T = T_1 + iT_2$, where T_1 and T_2 are Hermitian elements of $B(X)$.
(iii) T is a normal element of $\mathcal{A}_\ell(X)$ if and only if T^{**} is a normal element of $\mathcal{A}_\ell(X^{**})$.

PROOF. (i) We have proven above that if $T \in \mathcal{A}_\ell(X)_{sa}$, then $T^{**} \in \mathcal{A}_\ell(X^{**})_{sa}$. Conversely, suppose that $T^{**} \in \mathcal{A}_\ell(X^{**})_{sa}$. Since $T^{**}(X) = T(X) \subset X$, $T = T^{**}|_X \in \mathcal{A}_\ell(X)_{sa}$ by Proposition 5.2.

(ii) ($\Rightarrow$) We proved above that if $T \in \mathcal{A}_\ell(X)$ then $T^{**} \in \mathcal{A}_\ell(X^{**})$. The other assertion follows since $\mathcal{A}_\ell(X)$ is a C^*-algebra, and is a unital subalgebra of $B(X)$.

($\Leftarrow$) Suppose that $T = T_1 + iT_2$, where T_1 and T_2 are Hermitian elements of $B(X)$. Then $T^{**} = T_1^{**} + iT_2^{**}$, where T_1^{**} and T_2^{**} are Hermitian elements of $B(X^{**})$. On the other hand, assuming that $T^{**} \in \mathcal{A}_\ell(X^{**})$, one has that $T^{**} = R + iS$, where $R, S \in \mathcal{A}_\ell(X^{**})_{sa}$. In particular, R and S are Hermitian elements of $B(X^{**})$. Thus, by Lemma A.9, $T_1^{**}, T_2^{**} \in \mathcal{A}_\ell(X^{**})_{sa}$. By what has already been shown, $T_1, T_2 \in \mathcal{A}_\ell(X)_{sa}$ and so $T \in \mathcal{A}_\ell(X)$.

(iii) We leave the ($\Rightarrow$) direction as an easy exercise. For the other direction, we will make use of an elegant argument due to Behrends [**Beh2**, Theorem 1], which uses a Banach algebra version of Fuglede's theorem: namely if h, k are commuting Hermitian elements in a unital Banach algebra B, and if b commutes with $h + ik$, then b commutes with $h - ik$. The proof of this is identical to the proof of Fuglede's theorem found in modern basic functional analysis texts.

Suppose that T^{**} is a normal element in $\mathcal{A}_\ell(X^{**})$, then $T^{**} = H + iK$, for two commuting selfadjoint elements of $\mathcal{A}_\ell(X^{**})$. By (ii), it suffices to show that H and K are 'second adjoints' of bounded maps on X (which will then necessarily also be Hermitian). Now H, K are weak-* continuous, by the line after Theorem 2.3. Let us write H_* and K_* for the associated maps on X^*. Thus, it suffices to show that H_* and K_* are weak-* continuous maps on X^*. Since $T^* = H_* + iK_*$, it suffices to show that $H_* - iK_*$ is weak-* continuous. However a map S on X^* is weak-* continuous if and only if $S^* \circ i_X$ maps into $i_X(X)$, where i_X is the canonical map from X to X^{**}. Rephrasing this via the Hahn-Banach theorem, S is weak-* continuous if and only if whenever $\sigma \in X^{***}$ with $(i_X)^*(\sigma) = 0$, then $S^{**}(\sigma) \circ i_X = (i_X)^*(S^{**}(\sigma)) = 0$. Let $P = i_{X^*} \circ (i_X)^*$, which is an idempotent map on X^{***}, since $(i_X)^* \circ i_{X^*}$ is the identity map. By the above, it is easy to argue that S is weak-* continuous if and only if S^{**} commutes with P. We wish to apply this principle, taking $S = H_* - iK_*$, where $T^* = H_* + iK_*$ as above. Since T^{***} commutes with P, so does S^{**}, by the version of Fuglede's theorem mentioned above. Thus S is weak-* continuous. □

Remark. From (ii) of the last result one may give a solution to a problem raised in [**Beh2**]. To describe the problem we first list some notation. Behrends defines a map R on a Banach space X to 'possess an adjoint' if there exists an operator S on X (necessarily unique) such that both $\frac{1}{2}(R + S)$ and $\frac{1}{2i}(R - S)$ are Hermitian in $B(X)$. We shall call such an S a 'Behrends adjoint of R', and write $S = R^\flat$. Behrends asks the following question: if T is a map on a Banach space, such that $T^* \in B(X^*)$ possesses an adjoint in his sense, then is this adjoint necessarily weak-* continuous? This is equivalent to asking whether T 'possesses a Behrends adjoint' if and only if T^* does. In fact, any linear map T on an operator space X with $T \notin \mathcal{A}_\ell(X)$ but $T^{**} \in \mathcal{A}_\ell(X^{**})$ (for example a left multiplier T on a C^*-algebra which is not a two-sided multiplier, by the remarks above Proposition 5.16) gives rise to a counterexample to the question. Indeed, given such a T, let V be the involution in $\mathcal{A}_\ell(X^{**})$ of T^{**}. We have that $\frac{1}{2}(T^{**} + V)$ and $\frac{1}{2i}(T^{**} - V)$ are selfadjoint in $\mathcal{A}_\ell(X^{**})$, and consequently Hermitian also in $B(X^{**})$. We know that V is weak-* continuous by a fact listed after Theorem 2.3. Hence $V = S^*$ for some $S \in B(X^*)$. Now $\frac{1}{2}(T^* + S)$ and $\frac{1}{2i}(T^* - S)$ are both Hermitian in $B(X^*)$, since their Banach space adjoints are Hermitian. So T^* 'possesses a Behrends adjoint': $(T^*)^\flat = S$. However S is not weak*-continuous. For if it was, and $S = R^*$ say, then $\frac{1}{2}(T + R)$ and $\frac{1}{2i}(T - R)$ would be Hermitian. Thus by the last Proposition, T would be left adjointable, which is a contradiction.

In general, there is no way to produce from an element of $\mathcal{A}_\ell(X^{**})$ a corresponding element of $\mathcal{A}_\ell(X)$. If, however, X is a dual space, then there is. More precisely, we have the following result:

THEOREM 5.17. *Let X be an operator space. Then there exists a normal (i.e. weak-* continuous) conditional expectation E of $\mathcal{A}_\ell(X^{***})$ onto $\mathcal{A}_\ell(X^*)$. The weak-* topologies on these spaces are the canonical ones with respect to which they are W^*-algebras.*

PROOF. For any operator space Y, denote by $\iota_Y : Y \to Y^{**}$ the canonical completely isometric inclusion. We define $E : CB(X^{***}) \to CB(X^*)$ by the formula

$$E(T) = \iota_X^* \circ T \circ \iota_{X^*}.$$

Obviously, E is unital and contractive. Now suppose that $T \in \mathcal{M}_\ell(X^{***})$, with $\|T\|_{\mathcal{M}_\ell(X^{***})} \leq 1$. Then (claim) $E(T) \in \mathcal{M}_\ell(X^*)$, with $\|E(T)\|_{\mathcal{M}_\ell(X^*)} \leq 1$. Indeed, since $\|\tau_T^c\|_{cb} \leq 1$, one has that

$$\left\| \begin{bmatrix} E(T)(f) \\ g \end{bmatrix} \right\| = \left\| \begin{bmatrix} \iota_X^*(T(\iota_{X^*}(f))) \\ \iota_X^*(\iota_{X^*}(g)) \end{bmatrix} \right\| \leq \left\| \begin{bmatrix} T(\iota_{X^*}(f)) \\ \iota_{X^*}(g) \end{bmatrix} \right\| \leq \left\| \begin{bmatrix} \iota_{X^*}(f) \\ \iota_{X^*}(g) \end{bmatrix} \right\| = \left\| \begin{bmatrix} f \\ g \end{bmatrix} \right\|$$

for all $f, g \in X^*$, which says that $\|\tau_{E(T)}^c\| \leq 1$. Repeating this calculation for matrices over X^* shows that in fact $\|\tau_{E(T)}^c\|_{cb} \leq 1$, which in turn implies that $E(T) \in \mathcal{M}_\ell(X^*)$, with $\|E(T)\|_{\mathcal{M}_\ell(X^*)} \leq 1$. Since $E : \mathcal{M}_\ell(X^{***}) \to \mathcal{M}_\ell(X^*)$ is unital and contractive, it restricts to a positive map from $\mathcal{A}_\ell(X^{***})$ to $\mathcal{A}_\ell(X^*)$. The map $E : \mathcal{A}_\ell(X^{***}) \to \mathcal{A}_\ell(X^*)$ is surjective, since

$$E(T^{**}) = \iota_X^* \circ T^{**} \circ \iota_{X^*} = \iota_X^* \circ \iota_{X^*} \circ T = T$$

for all $T \in \mathcal{A}_\ell(X^*)$. Similarly, we have

$$E(S \circ T^{**}) = \iota_X^* \circ S \circ T^{**} \circ \iota_{X^*} = \iota_X^* \circ S \circ \iota_{X^*} \circ T = E(S) \circ T$$

for all $S \in \mathcal{A}_\ell(X^{***})$ and all $T \in \mathcal{A}_\ell(X^*)$. Taking involutions we also have $E(T^{**} \circ S) = T \circ E(S)$, so that E is a conditional expectation. Finally, we claim that it is normal. Indeed, if (T_i) is a bounded net in $\mathcal{A}_\ell(X^{***})$, $T \in \mathcal{A}_\ell(X^{***})$, and $T_i \to T$ weak-*, then for any $f \in X^*$, we have that

$$T_i(\iota_{X^*}(f)) \to T(\iota_{X^*}(f))$$

with respect to $\sigma(X^{***}, X^{**})$. This implies that

$$E(T_i)(f) = \iota_X^*(T_i(\iota_{X^*}(f))) \to \iota_X^*(T(\iota_{X^*}(f))) = E(T)(f)$$

with respect to $\sigma(X^*, X)$. Since the choice of f was arbitrary, $E(T_i) \to E(T)$ in the weak-* topology of $\mathcal{A}_\ell(X^*)$. $\square$

COROLLARY 5.18. Let X be an operator space and J be a right M-ideal of X^*. Then $\overline{J}^{\mathrm{wk}^*}$, the closure of J in the weak-* topology of X^*, is a right M-summand of X^*.

PROOF. Let $P \in \mathcal{A}_\ell(X^{***})$ be the left M-projection onto $J^{\perp\perp}$. Set $Q = E(P) \in \mathcal{A}_\ell(X^*)$. We claim that Q is a left M-projection with range $\overline{J}^{\mathrm{wk}^*}$. Clearly, Q is a contraction in $\mathcal{A}_\ell(X^*)$, and it is positive as an element of that C^*-algebra since E is a positive map. Now let $f \in J$. Then $\iota_{X^*}(f) \in J^{\perp\perp}$, which implies that

$$Q(f) = \iota_X^*(P(\iota_{X^*}(f))) = \iota_X^*(\iota_{X^*}(f)) = f.$$

Since Q is weak-* continuous, as noted after Theorem 2.3, we have $Q(f) = f$ for all $f \in \overline{J}^{\mathrm{wk}^*}$. Since

$$Q(X^*) = \iota_X^*(P(\iota_{X^*}(X^*))) \subset \iota_X^*(J^{\perp\perp}) = \iota_X^*(\overline{\iota_{X^*}(J)}^{\mathrm{wk}^*}) \subset \overline{\iota_X^*(\iota_{X^*}(J))}^{\mathrm{wk}^*} = \overline{J}^{\mathrm{wk}^*},$$

we deduce that $\mathrm{ran}(Q) = \overline{J}^{\mathrm{wk}^*}$ and $Q^2 = Q$, which proves the claim. $\square$

COROLLARY 5.19. Let X be an operator space and J be a right M-ideal of X^*. Then $^\perp J$ is a left L-summand of X.

PROOF. $^\perp J$ is a closed linear subspace of X whose annihilator is a right M-summand of X^*:

$$(^\perp J)^\perp = \overline{J}^{\mathrm{wk}^*}.$$

$\square$

COROLLARY 5.20. A dual operator space with no nontrivial right M-summands has only weak-* dense, or trivial, right M-ideals.

COROLLARY 5.21. Let X be an operator space and J be a right M-ideal of X^*. If $x \in \overline{J}^{\mathrm{wk}^*}$ (the closure of J in the weak-* topology of X^*), then there exists a net (x_i) in J, with $\|x_i\| \leq \|x\|$ for all i, such that $x_i \to x$ weak-*.

PROOF. Let $P \in \mathcal{A}_\ell(X^{***})$ be the left M-projection onto $J^{\perp\perp}$, and $Q = E(P) \in \mathcal{A}_\ell(X^*)$ be the left M-projection onto $\overline{J}^{\mathrm{wk}^*}$. Set $\hat{x} = \iota_{X^*}(x) \in X^{***}$. Then $P\hat{x} \in J^{\perp\perp} \cong J^{**}$. Thus, there exists a net $(x_i) \in J$, with $\|x_i\| \leq \|P\hat{x}\| \leq \|\hat{x}\| = \|x\|$ for all i, such that $\iota_{X^*}(x_i) \to P\hat{x}$ weak-*. But then $x_i = Qx_i = \iota_X^*(P(\iota_{X^*}(x_i))) \to \iota_X^*(P(\hat{x})) = Qx = x$ weak-*. $\square$

Remarks. 1) The classical analogs of the last four results are true too, and follow as a special case of the above. We have not seen these results in the literature, though.

2) A good illustration of Corollary 5.20 is $H^\infty(D)$, the classical H^∞ space of the unit disk. The M-projections on $H^\infty(D)$ (which are the same as the one-sided M-projections by facts in Section 4.2) are multiplications by characteristic functions in $H^\infty(D)$, and hence are trivial. Thus by Corollary 5.20 the M-ideals (which coincide with the one-sided M-ideals by 4.2, and which are known to be the closed

ideals of $H^\infty(D)$ possessing a bounded approximate identity (see e.g. [**HWW**]))
are either trivial or weak-* dense. In this spirit, note that the disk algebra $A(D)$
also has no nontrivial M-projections, but has plenty of M-ideals.

3) Another interesting example, related to the last remark, comes from work
of Powers [**Pow**]. Namely, there exists a C^*-algebra with no nontrivial M-ideals,
but with uncountably many pairwise non-isometric M-summands in the second
dual [**HWW**, Proposition V.4.6]. It is easy to show using basic spectral theory,
however, that every nontrivial C^*-algebra has nontrivial closed left ideals. On the
other hand, there are plenty of nonselfadjoint unital operator algebras with no
nontrivial left M-ideals. In fact it is easy to find two- or three-dimensional 'upper
triangular' matrix algebras with this property.

4) One may use some of the techniques above to construct a kind of polar
decomposition for a left adjointable multiplier T in a general operator space X.
The idea is to apply Theorem 3.13 together with Corollary 5.15 to the map $T^{**} \in
\mathcal{A}_\ell(X^{**})$. We obtain a partial isometry $W \in \mathcal{A}_\ell(X^{**})$ such that $T^{**} = W|T|^{**}$
and $|T|^{**} = W^\star T^{**}$. Restricting to X we obtain $T = V|T|$ and $|T| = UT$, where
$V = W|_X$ and $U = W^*|_X$. Unfortunately V (resp. U) maps into X^{**}, although on
$\mathrm{ran}(|T|)$ (resp. $\mathrm{ran}(T)$) it maps completely isometrically into X.

5) Corollary 5.21 is a 'Kaplansky density theorem' for right M-ideals in a dual
operator space.

6) Proposition 5.16 (iii) also seems to yield new results when applied to specific
examples. It follows, for example, when applied to a linear map T on a C^*-module,
that if T^{**} is adjointable as a map on X^{**}, and normal, then T is adjointable as a
map on X. (The 'normal' condition here cannot be dropped, by the example above
that proposition.)

We now turn to the implications of some of the results above to the Cunningham
algebra. As we saw in Section 2.4, the linear isometry $CB(X)^{\mathrm{op}} \to CB(X^*)$:
$T \mapsto T^*$ restricts to a *-isomorphism $\mathcal{C}_\mathrm{r}(X) \to \mathcal{A}_\ell(X^*)$ which is also a weak-
* homeomorphism. We claim that the corresponding linear isometry $CB(X) \to
CB(X^*)^{\mathrm{op}}$ restricts to a *-homomorphism $\mathcal{A}_\ell(X) \to \mathcal{C}_\mathrm{r}(X^*)$. Our argument is
indirect: we saw above that the linear isometry $\Phi : CB(X) \to CB(X^{**}) : T \to T^{**}$
restricts to a *-homomorphism $\mathcal{A}_\ell(X) \to \mathcal{A}_\ell(X^{**})$, which then can be composed
with the inverse of the *-isomorphism $\mathcal{C}_\mathrm{r}(X^*) \to \mathcal{A}_\ell(X^{**})$. It is interesting that a
direct argument does not present itself. Indeed, since the generic element of $\mathcal{A}_\ell(X)$
is not a norm limit of linear combinations of left M-projections on X, the following
result is not a priori clear:

COROLLARY 5.22. *The Banach space adjoint of an element of $\mathcal{A}_\ell(X)$ is an
element of $\mathcal{C}_\mathrm{r}(X^*)$.*

By composing the *-isomorphism $\mathcal{C}_\mathrm{r}(X) \to \mathcal{A}_\ell(X^*)$ with the *-homomorphism
$\mathcal{A}_\ell(X^*) \to \mathcal{C}_\mathrm{r}(X^{**})$, we obtain a *-homomorphism $\mathcal{C}_\mathrm{r}(X) \to \mathcal{C}_\mathrm{r}(X^{**}) : T \mapsto T^{**}$.

Obviously, there is a normal conditional expectation $\tilde{E} : \mathcal{C}_\mathrm{r}(X^{**}) \to \mathcal{C}_\mathrm{r}(X)$
which makes the following diagram commute:

$$
\begin{array}{ccc}
\mathcal{A}_\ell(X^{***}) & \xrightarrow{\ E\ } & \mathcal{A}_\ell(X^*) \\
{\scriptstyle T \mapsto T^*}\big\uparrow & & \big\uparrow{\scriptstyle T \mapsto T^*} \\
\mathcal{C}_\mathrm{r}(X^{**}) & \xrightarrow[\ \tilde{E}\ ]{} & \mathcal{C}_\mathrm{r}(X)
\end{array}
$$

Despite this, and in contrast to Corollary 5.18, it is not true that the weak-* closure of a left L-summand in a dual space is also a left L-summand. To see this we will use the fact that the intersection $J \cap K$ of two right M-summands J and K need not be a right M-ideal (see Example 5.28). Hence $(J \cap K)^\perp$, which by Appendix A.3 (ii) equals $\overline{J^\perp + K^\perp}^{\text{wk*}}$, is not a left L-summand. But the last set is the weak-* closure of $J^\perp + K^\perp$, which is a left L-summand by Corollary 3.4.

5.4. Sum and Intersection

In this section we describe how one-sided M-summands, M-ideals, and L-summands behave with respect to sums and intersections. Generally speaking, the behavior with respect to sums is better than the behavior with respect to intersections. Our first result follows immediately from Proposition 3.5.

PROPOSITION 5.23.   (i) If X is a dual operator space and $\{J_a : a \in A\}$ is a family of right M-summands of X, then $\cap_{a \in A} J_a$ and $\overline{\text{span}}^{\text{wk*}} \{\cup_{a \in A} J_a\}$ are right M-summands of X.

(ii) If X is an operator space and $\{J_a : a \in A\}$ is a family of left L-summands of X, then $\cap_{a \in A} J_a$ and $\overline{\text{span}} \{\cup_{a \in A} J_a\}$ are left L-summands of X.

COROLLARY 5.24. Let X be an operator space and $\{J_a : a \in A\}$ be a family of right M-ideals of X. Then $\overline{\text{span}} \{\cup_{a \in A} J_a\}$ is a right M-ideal of X.

PROOF. By Lemma A.3 and Proposition 5.23, $\overline{\text{span}} \{\cup_{a \in A} J_a\}$ is a closed linear subspace of X whose annihilator, $\cap_{a \in A} J_a^\perp$, is a left L-summand of X^*. $\square$

As the following two examples show, the 'closure' is necessary in both Proposition 5.23 and Corollary 5.24.

EXAMPLE 5.25. Let H be a Hilbert space, and J_1 and J_2 be closed linear subspaces of H such that $J_1 + J_2$ is not closed (cf. [**Hal**], §15). Let $X = H_c$, the column Hilbert space corresponding to H. Then J_1 and J_2 are right M-summands of X (cf. [**BEZ**], Proposition 6.10), but $J_1 + J_2$ is not a right M-ideal of X. Consequently, the norm closure is necessary in Corollary 5.24.

EXAMPLE 5.26. There exists a von Neumann algebra $\mathcal{M}$ and projections $P, Q \in \mathcal{M}$ such that $\overline{P\mathcal{M} + Q\mathcal{M}}$ (norm closure) is not equal to $R\mathcal{M}$ for any projection $R \in \mathcal{M}$. Thus, there exist right M-summands J_1 and J_2 of a dual operator space X such that $\overline{J_1 + J_2}$ is not a right M-summand of X. Consequently, the weak-* closure is necessary in Proposition 5.23.

PROOF. Let H be a Hilbert space, and H_1 and H_2 be closed linear subspaces of H such that $H_1 + H_2$ is not closed. We may assume that $H = \overline{H_1 + H_2}$. Let $P \in B(H)$ be the projection onto H_1 and let $Q \in B(H)$ be the projection onto H_2. Let $\mathcal{M} \subset B(H)$ be any von Neumann algebra containing P and Q. Suppose that $\overline{P\mathcal{M} + Q\mathcal{M}} = R\mathcal{M}$ for some projection $R \in \mathcal{M}$. Then $RP = P$ and $RQ = Q$, so that $R = I$. Thus for any $0 < \varepsilon < 1$, there exist $S, T \in \mathcal{M}$ such that $\|I - PS - QT\| \leq \varepsilon$. But then $PS + QT$ is invertible, which implies that $H = \text{ran}(PS + QT) \subset \text{ran}(P) + \text{ran}(Q) = H_1 + H_2$, a contradiction. $\square$

Of course, in the presence of commutativity, the need for the closure disappears. Namely,

PROPOSITION 5.27. Let X be an operator space.

(i) If J and K are right M-summands of X (resp. left L-summands of X) whose corresponding left M-projections (resp. right L-projections) commute, then $J + K$ is a right M-summand of X (resp. left L-summand of X). In particular, $J + K$ is closed.

(ii) If J and K are right M-ideals of X, then $J + K$ is a right M-ideal of X if and only if $J + K$ is closed. This happens, for example, when the left M-projections corresponding to $J^{\perp\perp}$ and $K^{\perp\perp}$ (resp. the right L-projections corresponding to $J^{\perp}$ and $K^{\perp}$) commute.

PROOF. (i) Suppose P corresponds to J and Q corresponds to K. Then $P + Q - PQ$ corresponds to $J + K$.

(ii) The first assertion follows from Corollary 5.24. The second assertion follows from (i) above, with the help of Lemma A.5. $\square$

As the next example shows, the omission in Corollary 5.24 of a statement concerning the intersection of right M-ideals was not an oversight—no general statement can be made. The lack of closure under arbitrary intersections of M-ideals is a feature of the classical M-ideal theory (cf. [**HWW**], Example II.5.5). However, the lack of closure under finite intersections is a new feature of the one-sided theory.

EXAMPLE 5.28. There exists a unital operator algebra $\mathcal{B}$ and projections $\tilde{P}, \tilde{Q} \in \mathcal{B}$ such that $\tilde{P}\mathcal{B} \cap \tilde{Q}\mathcal{B}$ has no contractive left approximate identity. Thus, by facts mentioned in Section 4.5, the intersection of two right M-summands of an operator space X need not be a right M-ideal of X, let alone a right M-summand. Consequently, the intersection of two right M-ideals of an operator space need not be a right M-ideal.

PROOF. Let $\mathcal{A} \subset B(H)$ be the unital C^*-algebra constructed in Example 7.1 of [**BSZ**], the important features of which we now recall:

- $\mathcal{A}$ contains projections P and Q such that $\mathrm{ran}(P \wedge Q)$ is separable with orthonormal basis $\{x_1, x_2, ...\}$, and $P \wedge Q \notin \mathcal{A}$.
- The one-dimensional projection E_n with range $\mathrm{span}\{x_n\}$ is an element of $\mathcal{A}$ for all $n \in \mathbb{N}$.

Set $\mathcal{B} = R_\infty^w(C_\infty(\mathcal{A}))$. Then $\mathcal{B}$ is a unital operator algebra (cf. Appendix B). Set $\tilde{P} = P \otimes I_\infty$ and $\tilde{Q} = Q \otimes I_\infty$. Clearly, $\tilde{P}$ and $\tilde{Q}$ are projections in $\mathcal{B}$. Suppose $\{F_\lambda : \lambda \in \Lambda\}$ is a contractive left approximate identity for

$$J = \tilde{P}\mathcal{B} \cap \tilde{Q}\mathcal{B} = R_\infty^w(C_\infty(P\mathcal{A})) \cap R_\infty^w(C_\infty(Q\mathcal{A})) = R_\infty^w(C_\infty(P\mathcal{A} \cap Q\mathcal{A})).$$

Then

$$F_\lambda \begin{bmatrix} E_1 & E_2 & E_3 & \dots \\ 0 & 0 & 0 & \dots \\ 0 & 0 & 0 & \dots \\ \vdots & \vdots & \vdots & \ddots \end{bmatrix} \rightarrow \begin{bmatrix} E_1 & E_2 & E_3 & \dots \\ 0 & 0 & 0 & \dots \\ 0 & 0 & 0 & \dots \\ \vdots & \vdots & \vdots & \ddots \end{bmatrix}.$$

We may view this as a convergent net in $M_\infty^w(B(H))$. Multiplying on the right by

$$\begin{bmatrix} E_1 & 0 & 0 & \dots \\ E_2 & 0 & 0 & \dots \\ E_3 & 0 & 0 & \dots \\ \vdots & \vdots & \vdots & \ddots \end{bmatrix},$$

we see that

$$F_\lambda \begin{bmatrix} P \wedge Q & 0 & 0 & \cdots \\ 0 & 0 & 0 & \cdots \\ 0 & 0 & 0 & \cdots \\ \vdots & \vdots & \vdots & \ddots \end{bmatrix} \rightarrow \begin{bmatrix} P \wedge Q & 0 & 0 & \cdots \\ 0 & 0 & 0 & \cdots \\ 0 & 0 & 0 & \cdots \\ \vdots & \vdots & \vdots & \ddots \end{bmatrix}.$$

From this it follows that $F_\lambda(1,1)(P \wedge Q) \rightarrow (P \wedge Q)$. Using Lemma 2.9 in [**B5**], we deduce that $F_\lambda(1,1)^* = F_\lambda(1,1)^*(P \wedge Q) \rightarrow (P \wedge Q)$. Thus $P \wedge Q \in \mathcal{A}$, which is a contradiction.   □

There are a number of ways to alleviate this lack of closure under finite intersections. They all hinge on the following lemma.

LEMMA 5.29. *Let X be an operator space, and J and K be right M-ideals of X. If $J + K$ is norm closed (equivalently if $J^\perp + K^\perp$ is norm closed or weak-* closed), then $J \cap K$ is a right M-ideal of X.*

PROOF. By Lemma A.5, the fact that $J + K$ is norm closed is equivalent to the fact that $J^\perp + K^\perp$ is norm closed, which in turn is equivalent to the fact that $J^\perp + K^\perp$ is weak-* closed. Now suppose that $J + K$ is norm closed. Then by Lemma A.3,

$$(J \cap K)^\perp = \overline{J^\perp + K^\perp}^{\mathrm{wk}^*} = J^\perp + K^\perp = \overline{J^\perp + K^\perp},$$

which is a left L-summand of X^* by Proposition 5.23.   □

The following results are stated for two right M-ideals; we leave the case of a finite number of right M-ideals as an exercise.

One situation when we can apply Lemma 5.29 is when there is 'commutativity'. Namely,

PROPOSITION 5.30. *Let X be an operator space.*
  (i) *If J and K are right M-summands of X whose corresponding left M-projections commute, then $J \cap K$ is a right M-summand of X.*
  (ii) *If J and K are right M-ideals of X such that the left M-projections corresponding to $J^{\perp\perp}$ and $K^{\perp\perp}$ (resp. the right L-projections corresponding to $J^\perp$ and $K^\perp$) commute, then $J \cap K$ is a right M-ideal of X.*

PROOF. (i) Let P and Q be the left M-projections of X corresponding to J and K, respectively. Then PQ is the left M-projection of X corresponding to $J \cap K$.

(ii) This follows immediately from Proposition 5.27 and Lemma 5.29.   □

A second situation when we can apply Lemma 5.29 is when there is 'finite-dimensionality'. Namely,

COROLLARY 5.31. *Let X be an operator space, and J and K be right M-ideals of X. If J is finite-dimensional or finite-codimensional, then $J \cap K$ and $J + K$ are right M-ideals of X.*

PROOF. Suppose J is finite-dimensional. Then $J + K$ is closed (Lemma A.8). Thus $J \cap K$ is a right M-ideal of X by Lemma 5.29, and $J + K$ is a right M-ideal by Proposition 5.27 (ii). Indeed since $J \cap K$ is finite-dimensional, it is a right M-summand of X (by a fact stated towards the end of §2.3). Now suppose that J is finite-codimensional. Then $J^\perp \cong (X/J)^*$ is finite-dimensional, so that $J^\perp + K^\perp$ is closed. Again we appeal to Lemma 5.29 and Proposition 5.27.   □

A third situation when we can apply Lemma 5.29 is when there is 'perpendicularity':

COROLLARY 5.32. Let X be an operator space, and let J and K be right M-ideals of X. If $J \curlywedge K$ then $J + K$ and $J \cap K$ are right M-ideals of X. Of course, $J \cap K = \{0\}$ in this case.

PROOF. There exists a completely isometric embedding σ of X into a C^*-algebra satisfying (3.2) for all $x \in J, y \in K$. Then σ^{**} is a completely isometric embedding of X^{**} into a W^*-algebra. Since σ^{**} is weak-* continuous, it is easy to check that σ^{**} satisfies the analog of (3.2) for all $x \in J^{\perp\perp}, y \in K^{\perp\perp}$. Let P and Q be the left M-projections onto $J^{\perp\perp}$ and $K^{\perp\perp}$, respectively. If $\langle \cdot, \cdot \rangle$ is the 'left Shilov inner product' discussed in Section 3.4, then we have by an observation there that

$$\langle QPx, y \rangle = \langle Px, Qy \rangle = 0,$$

for all $x, y \in X^{**}$. Thus $QP = 0 = PQ$, and we may appeal to Propositions 5.27 and 5.30. □

A final situation when we can apply Lemma 5.29 is when there is 'obliqueness', in a sense analogous to the 'positive angle' condition of Akemann (see [**Ake1**, Theorem II.7]). Namely,

PROPOSITION 5.33. Let X be an operator space, and let J and K be right M-ideals of X. Let P and Q be the left M-projections corresponding to $J^{\perp\perp}$ and $K^{\perp\perp}$ (resp. the right L-projections corresponding to $J^\perp$ and $K^\perp$). If $\|P(Q - P \wedge Q)\| < 1$, then $J \cap K$ and $J + K$ are right M-ideals of X.

Before proceeding with the proof, we interpret the condition appearing in the proposition. Consider orthogonal projections P and Q on a Hilbert space H. Let J and K be the ranges of P and Q, respectively. Then one can easily show that $\|PQ\| < 1$ if and only if there exists $0 \le \alpha < 1$ such that $|\langle x, y \rangle| \le \alpha$ for all $x \in \text{ball}(J)$ and all $y \in \text{ball}(K)$. In particular, $J \cap K = \{0\}$. More generally, $\|P(Q - P \wedge Q)\| < 1$ if and only if there exists $0 \le \alpha < 1$ such that $|\langle x, y \rangle| \le \alpha$ for all $x \in \text{ball}(J \cap (J \cap K)^\perp)$ and all $y \in \text{ball}(K \cap (J \cap K)^\perp)$. So as long as neither J nor K is contained in the other, the condition $\|P(Q - P \wedge Q)\| < 1$ expresses the fact that the 'angle' between J and K is positive (in the Hilbert space case, at least).

Now we come to the proof of the proposition.

PROOF. By Lemma 5.29, it suffices to prove that $\text{ran}(P) + \text{ran}(Q)$ is closed. By Lemma A.10, $\text{ran}(P) + \text{ran}(Q - P \wedge Q)$ is closed. But (claim) $\text{ran}(P) + \text{ran}(Q) = \text{ran}(P) + \text{ran}(Q - P \wedge Q)$. Indeed,

$$Px + Qy = P(x + (P \wedge Q)y) + (Q - P \wedge Q)y$$

and

$$Px + (Q - P \wedge Q)y = Px + Q(y - (P \wedge Q)y).$$

□

Remarks. 1) Proposition 5.33 is a more general form of Theorem II.7 of [**Ake1**], and indeed our result was inspired by that Theorem. We thank L. G. Brown for drawing our attention to this result of Akemann.

2) Proposition 5.33 contains Proposition 5.30 and Corollary 5.31 as special cases, a fact which the energetic reader can confirm.

One question remains here: Is the intersection of two right M-ideals of X^* again a right M-ideal of X^*? We doubt it, but no counter-example comes to mind.

5.5. Algebraic Direct Sum

We turn to another item in the classical M-ideal 'calculus'. Namely, if X is a Banach space, and J and K are M-ideals of X such that $X = J \oplus K$ (internal algebraic direct sum), then J and K are in fact complementary M-summands, i.e. $X = J \oplus_\infty K$ (see [**Beh1**], Proposition 2.8). Furthermore, an M-ideal J of a Banach space X is an M-summand of X if and only if there exists an M-ideal K of X such that $X = J \oplus K$ (same reference). The corresponding statements are not true for operator spaces and one-sided M-ideals.

EXAMPLE 5.34. Let $\mathcal{A} = C[0,1]$ and $\mathcal{I} = \{h \in C[0,1] : h(0) = 0\}$. Define $X = \mathcal{A} \oplus \mathcal{I}$ (external direct sum). Then X is a left Hilbert C^*-module over $\mathcal{A}$. Set

$$J = \{(h, h) : h \in \mathcal{I}\} \text{ and } K = \{(f, 0) : f \in \mathcal{A}\}.$$

Then J and K are closed left submodules (i.e. left M-ideals) of X such that $X = J \oplus K$ (internal direct sum). However, J is not orthogonally complemented (i.e. not a left M-summand). Indeed, if $\langle (f,g), (h,h) \rangle = 0$ for all $h \in \mathcal{I}$, then $(f+g)\overline{h} = 0$ for all $h \in \mathcal{I}$. This yields $f + g = 0$ on $(0,1]$, which in turn yields that $f + g = 0$. Thus, $f \in \mathcal{I}$ and $(f,g) \in \mathcal{I} \oplus \mathcal{I}$. It follows that not every element of X can be written as the sum of an element of J and an element orthogonal to J.

It is precisely 'commutativity' that is the missing ingredient (cf. Proposition 5.36 below). We begin with a simple lemma.

LEMMA 5.35. *Let X be an operator space, and J and K be closed linear subspaces of X. Then:*

(i) *J and K are complementary right M-summands of X if and only if $J^\perp$ and $K^\perp$ are complementary left L-summands of X^*.*

(ii) *J and K are complementary left L-summands of X if and only if $J^\perp$ and $K^\perp$ are complementary right M-summands of X^*.*

PROOF. (i) Suppose J and K are complementary right M-summands of X. Then there exists a left M-projection P on X such that $\operatorname{ran}(P) = J$ and $\ker(P) = K$. Then P^* is a right L-projection on X^* such that $\ker(P^*) = \operatorname{ran}(P)^\perp = J^\perp$ and $\operatorname{ran}(P^*) = \ker(I - P^*) = \operatorname{ran}(I - P)^\perp = \ker(P)^\perp = K^\perp$. Thus, $J^\perp$ and $K^\perp$ are complementary left L-summands of X^*. Conversely, suppose that $J^\perp$ and $K^\perp$ complementary left L-summands of X^*. Then there exists a right L-projection Q on X^* such that $\operatorname{ran}(Q) = J^\perp$ and $\ker(Q) = K^\perp$. By Lemma A.7, Q is weak-* continuous. Thus, there exists a bounded projection P on X such that $P^* = Q$. Clearly, P is a left M-projection. Furthermore,

$$\operatorname{ran}(P) = \ker(I - P) = {}^\perp \operatorname{ran}(I - Q) = {}^\perp \ker(Q) = {}^\perp(K^\perp) = \overline{K} = K,$$

and

$$\ker(P) = {}^\perp \operatorname{ran}(Q) = {}^\perp(J^\perp) = \overline{J} = J.$$

Thus, J and K are complementary right M-summands of X.

(ii) The proof of this assertion is similar to that of the previous one (easier, in fact). ☐

PROPOSITION 5.36. *Let X be an operator space, and J and K be right M-ideals of X such that $X = J \oplus K$. Then J and K are complementary right M-summands of X if and only if the left M-projections corresponding to $J^{\perp\perp}$ and $K^{\perp\perp}$ (resp. the right L-projections corresponding to $J^{\perp}$ and $K^{\perp}$) commute.*

PROOF. Let P and Q be the right L-projections corresponding to $J^{\perp}$ and $K^{\perp}$, respectively. Suppose that J and K are complementary right M-summands of X. Then by Lemma 5.35, $J^{\perp}$ and $K^{\perp}$ are complementary left L-summands of X^*. Thus, $Q = I - P$, so that $PQ = QP = 0$. Conversely, suppose that $PQ = QP$. Since $X = J \oplus K$, $X^* = J^{\perp} \oplus K^{\perp}$ (Lemma A.6). But then $PQ = QP$ has range $J^{\perp} \cap K^{\perp} = \{0\}$. Consequently, $P + Q = P + Q - PQ = I$. Thus $J^{\perp}$ and $K^{\perp}$ are complementary left L-summands of X^*. By Lemma 5.35, J and K are complementary right M-summands of X. ☐

COROLLARY 5.37. *Let X be an operator space, and J be a right M-ideal of X. Then J is a right M-summand of X if and only if there exists a right M-ideal K of X such that $X = J \oplus K$ and the left M-projections corresponding to $J^{\perp\perp}$ and $K^{\perp\perp}$ (resp. the right L-projections corresponding to $J^{\perp}$ and $K^{\perp}$) commute.*

5.6. One-Sided M-Summands in Tensor Products

We give a method to verify that if P is a left M-projection on an operator space X, then $P \otimes \mathrm{Id}_Z$ is a left M-projection on $X \otimes_\beta Z$, and that $P(X) \otimes_\beta Z$ is the corresponding right M-summand, for a fairly general class of 'tensor products' $\otimes_\beta$. Indeed, the argument sketched in the next paragraph applies to any tensor product $\otimes_\beta$ with the following three properties:

(1) $- \otimes_\beta \mathrm{Id}_Z$ is *functorial*. That is, if $T : X_1 \to X_2$ is completely contractive, then $T \otimes \mathrm{Id}_Z : X_1 \otimes_\beta Z \to X_2 \otimes_\beta Z$ is completely contractive.
(2) The canonical map $C_2(X) \otimes Z \to C_2(X \otimes Z)$ extends to a completely isometric isomorphism $C_2(X) \otimes_\beta Z \cong C_2(X \otimes_\beta Z)$.
(3) The span of elementary tensors $x \otimes z$ for $x \in X, z \in Z$ is dense in $X \otimes_\beta Z$.

Properties (1)–(3) certainly hold, for example, for the minimal (i.e. spatial), Haagerup, or extended Haagerup tensor products of operator spaces. The argument also applies to the module Haagerup tensor product over an operator algebra $\mathcal{A}$, provided P is also a right $\mathcal{A}$-module map.

Suppose then that P is a left M-projection on X. By Theorem 2.5 (v), this is equivalent to saying that the maps ν_P^c and μ_P^c mentioned there give a sequence of complete contractions

$$X \xrightarrow{\nu_P^c} C_2(X) \xrightarrow{\mu_P^c} X.$$

Applying hypothesis (1) above, we obtain an induced sequence of complete contractions

$$X \otimes_\beta Z \xrightarrow{\nu_P^c \otimes \mathrm{Id}_Z} C_2(X) \otimes_\beta Z \xrightarrow{\mu_P^c \otimes \mathrm{Id}_Z} X \otimes_\beta Z.$$

From (2) we obtain a sequence

$$X \otimes_\beta Z \longrightarrow C_2(X \otimes_\beta Z) \longrightarrow X \otimes_\beta Z.$$

From (3) it is easy to see that the first map in the last sequence equals $\nu_{P \otimes \mathrm{Id}_Z}^c$, while the second equals $\mu_{P \otimes \mathrm{Id}_Z}^c$. Thus these are complete contractions. Also, by

(3), $P \otimes \mathrm{Id}_Z$ is idempotent. Thus, by Theorem 2.5 (v), we deduce that $P \otimes \mathrm{Id}_Z$ is a left M-projection on $X \otimes_\beta Z$. By (1) again the canonical maps

$$P(X) \otimes_\beta Z \longrightarrow X \otimes_\beta Z \xrightarrow{P \otimes \mathrm{Id}_Z} P(X) \otimes_\beta Z$$

are complete contractions, which by (3) compose to the identity map. Hence the first map in the last sequence is a complete isometry. We conclude that $P(X) \otimes_\beta Z$ may be identified completely isometrically with a closed linear subspace of $X \otimes_\beta Z$. It is easy to see that this subspace is exactly the image of the left M-projection $P \otimes \mathrm{Id}_Z$.

In certain cases, a modification of the argument above will show that right M-ideals in X give rise to right M-ideals in $X \otimes_\beta Z$. The following proof is a good illustration of this technique.

THEOREM 5.38. *If J is a right M-ideal in an operator space X, then $J \otimes_\mathrm{h} Y$ is a right M-ideal in $X \otimes_\mathrm{h} Y$ for any operator space Y. Here $\otimes_\mathrm{h}$ is the Haagerup tensor product.*

PROOF. (Sketch) Suppose that P is the corresponding left M-projection from X^{**} onto $J^{\perp\perp}$. We need to show that $(J \otimes_h Y)^{\perp\perp}$ is a right M-summand in $(X \otimes_h Y)^{**}$. The latter space may be identified with the *normal Haagerup tensor product* $X^{**} \otimes_{\sigma h} Y^{**}$. See [**EK, ER3**] for the definition and basic properties of this tensor product. In particular it satisfies analogues of properties (1)–(3) above; namely

(1) $- \otimes_{\sigma h} \mathrm{Id}_{Y^{**}}$ is *functorial*. That is, if $T : X_1 \to X_2$ is weak-* continuous and completely contractive, then $T \otimes \mathrm{Id}_{Y^{**}} : X_1 \otimes_{\sigma h} Y^{**} \to X_2 \otimes_{\sigma h} Y^{**}$ is weak-* continuous and completely contractive.

(2) $C_2(X^{**}) \otimes_{\sigma h} Y^{**} \cong C_2(X^{**} \otimes_{\sigma h} Y^{**})$ completely isometrically and weak-* homeomorphically.

(3) The span of elementary tensors $x \otimes y$ for $x \in X, y \in Y$ is weak-* dense in $X^{**} \otimes_{\sigma h} Y^{**}$.

To see (2) for example, note that by basic properties of the Haagerup tensor product,

$$C_2(X^{**} \otimes_{\sigma h} Y^{**}) \cong C_2((X \otimes_h Y)^{**}) \cong C_2(X \otimes_h Y)^{**} \cong C_2(X^{**}) \otimes_{\sigma h} Y^{**},$$

and one can verify that each '$\cong$' here is a weak-* homeomorphism.

We may now follow the proof above Theorem 5.38, checking that all maps are weak-* continuous. We deduce that $P \otimes \mathrm{Id}_{Y^{**}}$ is a left M-projection on $X^{**} \otimes_{\sigma h} Y^{**}$, and its range is completely isometric and weak-* homeomorphic to $J^{\perp\perp} \otimes_{\sigma h} Y^{**}$. Since $P \otimes \mathrm{Id}_{Y^{**}}$ is the identity map when restricted to $J \otimes_h Y$, it is easy to see that the range of $P \otimes \mathrm{Id}_{Y^{**}}$ is $(J \otimes_h Y)^{\perp\perp} = \overline{J \otimes_h Y}^{\mathrm{wk}^*}$. $\square$

We now turn to different methods to analyze tensor products.

5.7. Minimal Tensor Product

We write $X \check{\otimes} Y$ for the minimal or spatial tensor product of operator spaces (see the standard operator space texts for more details).

A similar analysis to that in the last proof shows that in cases that $(X \check{\otimes} Y)^{**} \cong X^{**} \bar{\otimes} Y^{**}$ (see e.g. [**ER4**, Chapter 8] for this notation), if J is a right M-ideal in X then $J \check{\otimes} Y$ is a right M-ideal in $X \check{\otimes} Y$. We leave it to the reader to make this precise. Note however that $(X \check{\otimes} Y)^{**} \neq X^{**} \bar{\otimes} Y^{**}$ in many cases, even if X is finite dimensional—this is related to the topic of local reflexivity.

PROPOSITION 5.39. Let X and Y be operator spaces, $S \in \mathcal{M}_\ell(X)$, and $T \in \mathcal{M}_\ell(Y)$. Then $S \otimes T \in \mathcal{M}_\ell(X \check{\otimes} Y)$, and $\|S \otimes T\|_{\mathcal{M}_\ell(X \check{\otimes} Y)} \leq \|S\|_{\mathcal{M}_\ell(X)} \|T\|_{\mathcal{M}_\ell(Y)}$. If $S = \mathrm{Id}_X$ or $T = \mathrm{Id}_Y$, then in fact we have equality in the previous inequality. If $S \in \mathcal{A}_\ell(X)$ and $T \in \mathcal{A}_\ell(Y)$, then $S \otimes T \in \mathcal{A}_\ell(X \check{\otimes} Y)$ and $(S \otimes T)^\star = S^\star \otimes T^\star$.

PROOF. Let (σ, A) and (ρ, B) be implementing pairs for S and T, respectively. Then $\sigma : X \to B(H)$ and $\rho : Y \to B(K)$ are complete isometries. Thus, by the injectivity of the minimal tensor product, $\sigma \otimes \rho : X \check{\otimes} Y \to B(H) \check{\otimes} B(K) \subset B(H \otimes K)$ is a complete isometry. We claim that $(\sigma \otimes \rho, A \otimes B)$ is an implementing pair for $S \otimes T$. Indeed, for all $x \in X$ and $y \in Y$,

$$
\begin{aligned}
(\sigma \otimes \rho)((S \otimes T)(x \otimes y)) &= (\sigma \otimes \rho)(Sx \otimes Ty) = \sigma(Sx) \otimes \rho(Ty) \\
&= A\sigma(x) \otimes B\rho(y) = (A \otimes B)(\sigma(x) \otimes \rho(y)) \\
&= (A \otimes B)(\sigma \otimes \rho)(x \otimes y).
\end{aligned}
$$

It follows that $S \otimes T \in \mathcal{M}_\ell(X \check{\otimes} Y)$ and $\|S \otimes T\|_{\mathcal{M}_\ell(X \check{\otimes} Y)} \leq \|A \otimes B\| = \|A\|\|B\|$. Since the choice of implementing pairs was arbitrary,

$$
\|S \otimes T\|_{\mathcal{M}_\ell(X \check{\otimes} Y)} \leq \|S\|_{\mathcal{M}_\ell(X)} \|T\|_{\mathcal{M}_\ell(Y)}.
$$

Furthermore, if $A^*\sigma(X) \subset \sigma(X)$ and $B^*\rho(Y) \subset \rho(Y)$, then it is clear that we have $(A^* \otimes B^*)(\sigma \otimes \rho)(X \check{\otimes} Y) \subset (\sigma \otimes \rho)(X \check{\otimes} Y)$. Thus, if $S \in \mathcal{A}_\ell(X)$ and $T \in \mathcal{A}_\ell(Y)$, then $S \otimes T \in \mathcal{A}_\ell(X \check{\otimes} Y)$ and $(S \otimes T)^\star = S^\star \otimes T^\star$. Finally, suppose that $T = \mathrm{Id}_Y$. Let (θ, C) be an implementing pair for $S \otimes T$, so that $\theta : X \check{\otimes} Y \to B(L)$ is a complete isometry. Fix a unit vector $y \in Y$ and define $\theta_y : X \to B(L)$ by $\theta_y(x) = \theta(x \otimes y)$ for all $x \in X$. Then θ_y is a complete isometry and

$$
\theta_y(Sx) = \theta(Sx \otimes y) = \theta(Sx \otimes Ty) = C\theta(x \otimes y) = C\theta_y(x)
$$

for all $x \in X$. In other words, (θ_y, C) is an implementing pair for S. Thus, $\|S\|_{\mathcal{M}_\ell(X)} \|T\|_{\mathcal{M}_\ell(Y)} = \|S\|_{\mathcal{M}_\ell(X)} \leq \|C\|$. Since the choice of the implementing pair was arbitrary, $\|S\|_{\mathcal{M}_\ell(X)} \|T\|_{\mathcal{M}_\ell(Y)} \leq \|S \otimes T\|_{\mathcal{M}_\ell(X \check{\otimes} Y)}$. The same type of argument works if $S = \mathrm{Id}_X$. □

In particular, a left multiplier (resp. left adjointable multiplier) on an operator space X amplifies to a left multiplier (resp. left adjointable multiplier) on $M_I(X)$ or $K_I(X)$, for any index set I. The amplification has the same multiplier norm as the original map. In fact, the same is true for the amplification to $M_I^w(X)$, although this doesn't follow from the above proposition (it follows from a modification of the proof of Proposition 5.47 below; see also Corollary 5.49).

5.8. Haagerup Tensor Product

PROPOSITION 5.40. If T is a left multiplier on an operator space X, and if Y is any other operator space, then $T \otimes \mathrm{Id}_Y$ is a left multiplier on $X \otimes_h Y$, where $\otimes_h$ is the Haagerup tensor product. Furthermore, the multiplier norms of T and $T \otimes \mathrm{Id}_Y$ are the same. If T is a left adjointable multiplier on X, then $T \otimes \mathrm{Id}_Y$ is a left adjointable multiplier of $X \otimes_h Y$, and $(T \otimes \mathrm{Id}_Y)^\star = T^\star \otimes \mathrm{Id}_Y$.

PROOF. Suppose $T \in \mathcal{M}_\ell(X)$ and $\|T\|_{\mathcal{M}_\ell(X)} \leq 1$. Under the completely isometric identification $C_2(X \otimes_h Y) \cong C_2(X) \otimes_h Y$, $\tau^c_{T \otimes \mathrm{Id}_Y}$ corresponds to $\tau^c_T \otimes \mathrm{Id}_Y$. Therefore,

$$
\|\tau^c_{T \otimes \mathrm{Id}_Y}\|_{cb} = \|\tau^c_T \otimes \mathrm{Id}_Y\|_{cb} \leq \|\tau^c_T\|_{cb} \|\mathrm{Id}_Y\|_{cb} \leq 1,
$$

which says that $T \otimes \mathrm{Id}_Y \in \mathcal{M}_\ell(X \otimes_\mathrm{h} Y)$ and $\|T \otimes \mathrm{Id}_Y\|_{\mathcal{M}_\ell(X \otimes_\mathrm{h} Y)} \leq 1$. Now suppose that $T \otimes \mathrm{Id}_Y \in \mathcal{M}_\ell(X \otimes_\mathrm{h} Y)$. Let (σ, A) be an implementing pair for $T \otimes \mathrm{Id}_Y$. Fix a unit vector $y \in Y$ arbitrarily and define $\sigma_y : X \to B(H)$ by

$$\sigma_y(x) = \sigma(x \otimes y)$$

for all $x \in X$. Then σ_y is a complete isometry and

$$\sigma_y(Tx) = \sigma(Tx \otimes y) = \sigma((T \otimes \mathrm{Id}_Y)(x \otimes y)) = A\sigma(x \otimes y) = A\sigma_y(x)$$

for all $x \in X$. In other words, (σ_y, A) is an implementing pair for T. Thus, $\|T\|_{\mathcal{M}_\ell(X)} \leq \|A\|$. Since the choice of (σ, A) was arbitrary, $\|T\|_{\mathcal{M}_\ell(X)} \leq \|T \otimes \mathrm{Id}_Y\|_{\mathcal{M}_\ell(X \otimes_\mathrm{h} Y)}$. Now suppose that $T \in \mathcal{A}_\ell(X)_{sa}$. Then for all $t \in \mathbb{R}$,

$$\begin{aligned} \|\exp(it\tau^c_{T \otimes \mathrm{Id}_Y})\|_{cb} &= \|\exp(it(\tau^c_T \otimes \mathrm{Id}_Y))\|_{cb} \\ &= \|\exp(it\tau^c_T) \otimes \mathrm{Id}_Y\|_{cb} \leq \|\exp(it\tau^c_T)\|_{cb} = 1. \end{aligned}$$

Thus, $T \otimes \mathrm{Id}_Y \in \mathcal{A}_\ell(X \otimes_\mathrm{h} Y)_{sa}$. Finally, if $T \in \mathcal{A}_\ell(X)$, then $T = T_1 + iT_2$, where $T_1, T_2 \in \mathcal{A}_\ell(X)_{sa}$, so that

$$T \otimes \mathrm{Id}_Y = (T_1 \otimes \mathrm{Id}_Y) + i(T_2 \otimes \mathrm{Id}_Y) \in \mathcal{A}_\ell(X \otimes_\mathrm{h} Y).$$

Furthermore,

$$(T \otimes \mathrm{Id}_Y)^\star = (T_1 \otimes \mathrm{Id}_Y) - i(T_2 \otimes \mathrm{Id}_Y) = T^\star \otimes \mathrm{Id}_Y.$$

$\square$

Remarks. 1) Due to the lack of 'commutativity' in the Haagerup tensor product, $\mathrm{Id}_X \otimes T$ need not be a left multiplier (resp. left adjointable multiplier) of $X \otimes_\mathrm{h} Y$ when T is a left multiplier (resp. left adjointable multiplier) of Y. The correct statement is that $\mathrm{Id}_X \otimes T$ is a right multiplier (resp. right adjointable multiplier) of $X \otimes_\mathrm{h} Y$ when T is a right multiplier (resp. right adjointable multiplier) of Y.

2) Results similar to Proposition 5.40 hold for the extended and module Haagerup tensor products.

It follows immediately from Proposition 5.40 that the map $T \mapsto T \otimes \mathrm{Id}_Y$ is an isometric homomorphism from $\mathcal{M}_\ell(X)$ into $\mathcal{M}_\ell(X \otimes_\mathrm{h} Y)$, and that this map restricts to an injective $*$-homomorphism from $\mathcal{A}_\ell(X)$ into $\mathcal{A}_\ell(X \otimes_\mathrm{h} Y)$. Thus $\mathcal{A}_\ell(X)$ may be viewed as a $*$-subalgebra of $\mathcal{A}_\ell(X \otimes_\mathrm{h} Y)$. Typically this inclusion is proper. For example, $\mathcal{A}_\ell(C_n) = M_n$, but $\mathcal{A}_\ell(C_n \otimes_\mathrm{h} C_n) = \mathcal{A}_\ell(C_{n^2}) = M_{n^2}$. If $\mathcal{A}$ and $\mathcal{B}$ are nontrivial unital C^*-algebras, however, we have that $\mathcal{A}_\ell(\mathcal{A}) = \mathcal{A}_\ell(\mathcal{A} \otimes_\mathrm{h} \mathcal{B})$ (see Theorem 5.42 below). First we recall a well-known fact about the Haagerup tensor product, for which we include a proof for completeness.

LEMMA 5.41. *Let V and W be operator spaces and $x \in V \otimes_\mathrm{h} W$. If $(\phi \otimes \mathrm{Id}_W)(x) = 0$ for all $\phi \in V^*$, then $x = 0$. Likewise, if $(\mathrm{Id}_V \otimes \psi)(x) = 0$ for all $\psi \in W^*$, then $x = 0$.*

PROOF. We only prove the first assertion. By [**ER4**], Proposition 9.2.5 and Theorem 9.4.7, we have the completely isometric inclusions

$$V \otimes_\mathrm{h} W \to V^{**} \otimes_\mathrm{h} W^{**} \to (V^* \otimes_\mathrm{h} W^*)^*.$$

Let $\Phi_x \in (V^* \otimes_\mathrm{h} W^*)^*$ be the image of x under these inclusions. We have

$$\Phi_x(\phi \otimes \psi) = (\phi \otimes \psi)(x) = \psi((\phi \otimes \mathrm{Id}_W)(x)) = 0$$

for all $\phi \in V^*$ and all $\psi \in W^*$. It follows that $\Phi_x = 0$, which implies that $x = 0$. $\square$

THEOREM 5.42. *Let $\mathcal{A}$ and $\mathcal{B}$ be unital C^*-algebras, with neither $\mathcal{A}$ nor $\mathcal{B}$ equal to $\mathbb{C}$. Then $\mathcal{A} \cong \mathcal{A}_\ell(\mathcal{A} \otimes_{\mathrm{h}} \mathcal{B})$ and $\mathcal{B} \cong \mathcal{A}_{\mathrm{r}}(\mathcal{A} \otimes_{\mathrm{h}} \mathcal{B})$.*

PROOF. Let $\lambda : \mathcal{A} \to \mathcal{A}_\ell(\mathcal{A} \otimes_{\mathrm{h}} \mathcal{B})$ and $\rho : \mathcal{B} \to \mathcal{A}_{\mathrm{r}}(\mathcal{A} \otimes_{\mathrm{h}} \mathcal{B})$ be the canonical injective *-homomorphisms (see the discussion above Lemma 5.41). We aim to show that λ is surjective (the proof that ρ is surjective being similar). It suffices to show that any self-adjoint $T \in \mathcal{A}_\ell(\mathcal{A} \otimes_{\mathrm{h}} \mathcal{B})$, with $\|T\|_{\mathcal{M}_\ell(\mathcal{A} \otimes_{\mathrm{h}} \mathcal{B})} \leq 1$, is in the range of λ. We will use R. R. Smith's observation that $\mathcal{A} \otimes_{\mathrm{h}} \mathcal{B}$ is a unital Banach algebra with product

$$(a \otimes b)(a' \otimes b') = aa' \otimes bb'$$

(see e.g. Proposition 2 in [**B2**]). From Lemma 2.4 we know that $T(1 \otimes 1) \in \mathrm{Her}(\mathcal{A} \otimes_{\mathrm{h}} \mathcal{B})$. Thus by (1) on p. 126 of [**B2**] we have that

$$(5.1) \qquad T(1 \otimes 1) = h \otimes 1 + 1 \otimes k$$

for some $h \in \mathcal{A}_{sa}$, $k \in \mathcal{B}_{sa}$. Since left and right multipliers of an operator space automatically commute, $\rho(\mathcal{B})$ commutes with T. Thus

$$(5.2) \qquad T(a \otimes b) = T(\rho(b)(a \otimes 1)) = \rho(b)(T(a \otimes 1)) = T(a \otimes 1)(1 \otimes b)$$

for $a \in \mathcal{A}$, $b \in \mathcal{B}$. We next will prove the identity

$$(5.3) \qquad (\mathrm{Id}_\mathcal{A} \otimes \psi)(T(a \otimes w)) = (\mathrm{Id}_\mathcal{A} \otimes \psi)(T(1 \otimes w))a$$

for all $\psi \in \mathcal{B}^*$ and $w \in \mathcal{B}$. First suppose that w is a unitary in $\mathcal{B}$, and that $\psi \in \mathcal{B}^*$ satisfies $\psi(w) = 1 = \|\psi\|$. Consider the operator $u(a) = (\mathrm{Id}_\mathcal{A} \otimes \psi)(T(a \otimes w))$ on $\mathcal{A}$. We have for any $a' \in \mathcal{A}$ that

$$\left\| \begin{bmatrix} u(a) \\ a' \end{bmatrix} \right\| = \left\| \begin{bmatrix} (\mathrm{Id}_\mathcal{A} \otimes \psi)(T(a \otimes w)) \\ (\mathrm{Id}_\mathcal{A} \otimes \psi)(a' \otimes w) \end{bmatrix} \right\| \leq \left\| \begin{bmatrix} T(a \otimes w) \\ a' \otimes w \end{bmatrix} \right\| \leq \left\| \begin{bmatrix} a \otimes w \\ a' \otimes w \end{bmatrix} \right\|,$$

the last inequality by Theorem 2.1. Since we clearly have

$$\left\| \begin{bmatrix} a \otimes w \\ a' \otimes w \end{bmatrix} \right\| = \left\| \begin{bmatrix} a \\ a' \end{bmatrix} \right\|,$$

we see using Lemma 4.1 in [**BSZ**] that there exists an $a_{w,\psi} \in \mathcal{A}$ such that

$$(\mathrm{Id}_\mathcal{A} \otimes \psi)(T(a \otimes w)) = a_{w,\psi} a$$

for all $a \in \mathcal{A}$. Setting $a = 1$ gives $a_{w,\psi} = (\mathrm{Id}_\mathcal{A} \otimes \psi)(T(1 \otimes w))$, and this establishes (5.3) in this case.

If $g \in \mathcal{B}^*$, then $g_w = g(\cdot w) \in \mathcal{B}^*$. Since $\mathcal{B}^*$ is the span of the states on $\mathcal{B}$, we may write $g_w = \sum_{k=1}^4 \alpha_k f^k$ for α_k scalars and f^k states on $\mathcal{B}$. Then $g = \sum_{k=1}^4 \alpha_k f_{w^*}^k$. Setting $\psi = f_{w^*}^k$ in (5.3), and using the fact that Equation (5.3) is linear in ψ, we now have (5.3) with w unitary. By the well-known fact that the unitary elements span a C^*-algebra, and the linearity of Equation (5.3) in w, we obtain (5.3) for any $w \in \mathcal{B}$. Thus we have proved (5.3) in general.

Combining (5.3) and (5.2) we obtain

$$(\mathrm{Id}_\mathcal{A} \otimes \psi)(T(a \otimes b)) = (\mathrm{Id}_\mathcal{A} \otimes \psi)(T(1 \otimes 1)(1 \otimes b))a.$$

Writing $T(1 \otimes 1)$ as in (5.1), we have that

$$(\mathrm{Id}_\mathcal{A} \otimes \psi)(T(a \otimes b)) = \psi(b)ha + \psi(kb)a = (\mathrm{Id}_\mathcal{A} \otimes \psi)((h \otimes 1 + 1 \otimes k)(a \otimes b)).$$

Since ψ is arbitrary, we deduce by Lemma 5.42 that

$$T(a \otimes b) = (h \otimes 1 + 1 \otimes k)(a \otimes b)$$

for all $a \in \mathcal{A}$, $b \in \mathcal{B}$.

For any $x \in \mathcal{A} \otimes_{\mathrm{h}} \mathcal{B}$, denote by L_x the operator of left multiplication by x on $\mathcal{A} \otimes_{\mathrm{h}} \mathcal{B}$. Since $L_{h \otimes 1} = \lambda(h)$, $L_{h \otimes 1} \in \mathcal{A}_\ell(\mathcal{A} \otimes_{\mathrm{h}} \mathcal{B})_{sa}$. Since $T \in \mathcal{A}_\ell(\mathcal{A} \otimes_{\mathrm{h}} \mathcal{B})_{sa}$ too, we deduce that $L_{1 \otimes k} \in \mathcal{A}_\ell(\mathcal{A} \otimes_{\mathrm{h}} \mathcal{B})_{sa}$. We claim that k is a scalar multiple of 1. Suppose not. Let $c \in \mathcal{A}_{sa}$. Then $L_{c \otimes 1} = \lambda(c) \in \mathcal{A}_\ell(\mathcal{A} \otimes_{\mathrm{h}} \mathcal{B})_{sa}$ and $L_{c \otimes 1}$ commutes with $L_{1 \otimes k}$. It follows that $L_{c \otimes k} = L_{c \otimes 1} L_{1 \otimes k} \in \mathcal{A}_\ell(\mathcal{A} \otimes_{\mathrm{h}} \mathcal{B})_{sa}$. By Lemma 2.4,

$$c \otimes k = L_{c \otimes k}(1 \otimes 1) \in \mathrm{Her}(\mathcal{A} \otimes_{\mathrm{h}} \mathcal{B}),$$

and so by (1) on p. 126 of [**B2**],

$$c \otimes k = h_c \otimes 1 + 1 \otimes k_c$$

for some $h_c \in \mathcal{A}_{sa}$, $k_c \in \mathcal{B}_{sa}$. Now let $\phi \in \mathcal{A}^*$ be a state. Then

$$\phi(c)k = (\phi \otimes \mathrm{Id}_{\mathcal{B}})(c \otimes k) = (\phi \otimes \mathrm{Id}_{\mathcal{B}})(h_c \otimes 1 + 1 \otimes k_c) = \phi(h_c)1 + k_c,$$

so that

$$k_c = \phi(c)k - \phi(h_c)1.$$

But then

$$c \otimes k = h_c \otimes 1 + 1 \otimes \phi(c)k - 1 \otimes \phi(h_c)1,$$

which implies that

$$(c - \phi(c)1) \otimes k = (h_c - \phi(h_c)1) \otimes 1.$$

Since k and 1 are linearly independent, $c = \phi(c)1$. Since the choice of c was arbitrary, $\mathcal{A} = \mathbb{C}$, a contradiction. $\qquad\square$

COROLLARY 5.43. *Given any two unital C^*-algebras $\mathcal{A}$ and $\mathcal{B}$, there exists an operator space X with $\mathcal{A}_\ell(X) \cong \mathcal{A}$ and $\mathcal{A}_{\mathrm{r}}(X) \cong \mathcal{B}$. Given any two lattices L_1 and L_2 which are the projection lattices for two von Neumann algebras, there exists an operator space X whose lattice of left M-summands is lattice isomorphic to L_1 and whose lattice of right M-summands is lattice isomorphic to L_2.*

PROOF. If $\mathcal{A}$ and $\mathcal{B}$ are both nontrivial then this follows from the last Theorem immediately. So suppose that $\mathcal{A} = \mathbb{C}$ (we leave the remaining cases to the reader). Let $X = R_n \otimes_{\mathrm{h}} \mathcal{B}$. If $T \in \mathcal{A}_\ell(X)$, then by Proposition 5.39 we have that $\mathrm{Id}_{C_n} \otimes T \in \mathcal{A}_\ell(C_n \check{\otimes} X)$. By basic operator space theory we have

$$C_n \check{\otimes} X \cong C_n \otimes_{\mathrm{h}} R_n \otimes_{\mathrm{h}} \mathcal{B} \cong M_n \otimes_{\mathrm{h}} \mathcal{B}.$$

By the last Theorem it follows that there exists a matrix $A \in M_n$, with

$$x \otimes T(u) = Ax \otimes u$$

for all $x \in C_n, u \in X$. By linear algebra we may deduce that $T \in \mathbb{C} \mathrm{Id}_X$. Since the choice of T was arbitrary, $\mathcal{A}_\ell(X) = \mathbb{C}$. On the other hand, by Proposition 5.40,

$$\mathcal{B} \subset \mathcal{A}_{\mathrm{r}}(R_n \otimes_{\mathrm{h}} \mathcal{B}) \subset \mathcal{A}_{\mathrm{r}}(M_n \otimes_{\mathrm{h}} \mathcal{B}) = \mathcal{B}.$$

$\square$

5.9. Interpolation

In this section we shall see that the class of one-sided M-summands is closed under interpolation, and we will use this to construct some 'exotic' examples. We will use the interpolation formula $M_n(X_\theta) = M_n(X)_\theta$ and Proposition 2.1 from [**Pis2**]. Suppose that (E_0, E_1) is a compatible couple of operator spaces in the sense of interpolation theory. If $T : E_0 + E_1 \to E_0 + E_1$ restricts to a map $E_0 \to E_0$ which is a contractive left multiplier on E_0, and restricts to a map $E_1 \to E_1$ which is a contractive left multiplier on E_1, then we claim that T induces a left multiplier on E_θ, for all $\theta \in [0, 1]$. To see this note that $\tau_T^c : C_2(E_0) + C_2(E_1) \to C_2(E_0) + C_2(E_1)$ restricts to complete contractions $\tau_T^c : C_2(E_0) \to C_2(E_0)$ and $\tau_T^c : C_2(E_1) \to C_2(E_1)$. By [**Pis2**] Proposition 2.1, $\tau_T^c : C_2(E_\theta) \to C_2(E_\theta)$ is a complete contraction. Thus T has induced a canonical contractive left multiplier on each E_θ. Furthermore, if T is idempotent, then we have obtained a left M-projection on each E_θ, and have therefore interpolated between the left M-summands $T(E_0)$ and $T(E_1)$. It should be remarked that T need not be a left multiplier on $E_0 + E_1$. Indeed, τ_T^c is completely contractive on $C_2(E_0) + C_2(E_1)$, but not necessarily on $C_2(E_0 + E_1)$.

EXAMPLE 5.44. Let $E_0 = \ell^\infty$, and $E_1 = \ell^2$ with its column Hilbert space structure. View $\ell^2 \subset \ell^\infty$. Let I be a nontrivial subset of the natural numbers, and let F_0 and F_1 be the subspaces of E_0 and E_1 supported on I. These are right M-summands in E_0 and E_1 respectively. Interpolating between these spaces yields the somewhat surprising existence of right M-summands in ℓ^p (the latter Banach space equipped with some operator space structure) for $2 < p < \infty$. On the other hand it follows from (4) in [**BEZ**], in conjunction with Theorem 2.5 (iv), that for $1 \le p < 2$, there is no operator space structure on ℓ^p such that the sequences supported on I form a right M-summand.

EXAMPLE 5.45. A similar technique works for L^p-spaces and their noncommutative variant. In the commutative case note that $L^\infty(\Omega)$ acts as left multipliers both on $L^2(\Omega)_c$ and on $L^\infty(\Omega)$. Interpolating, we see that left multiplication by a function f in $L^\infty(\Omega)$ is a left multiplier on $L^p(\Omega)$ (the latter Banach space equipped with some operator space structure) for $2 < p < \infty$. Taking f to be a characteristic function yields left M-projections.

In the noncommutative case we do the same thing, after observing that a (finite) von Neumann algebra $\mathcal{M}$ acts as left multipliers both on the Hilbert space $L^2(\mathcal{M})_c$, and on $L^\infty(\mathcal{M}) = \mathcal{M}$. Interpolating we obtain exotic right M-summands and left multipliers on the Banach space $L^p(\mathcal{M})$ (with some operator space structure) for $2 < p < \infty$.

The operator space structure on $L^p(\mathcal{M})$ in the last example is not quite the standard one of [**Pis1**]. However it is not a very obscure structure. Marius Junge has informed us that he has encountered it in some of his work (see e.g. [**JP**]). In certain cases this structure has a simple description. For example if $\mathcal{M} = M_n$ then we are interpolating between M_n with its Hilbert-Schmidt norm and column Hilbert space structure, and M_n. That is, in operator space notation, we are interpolating between $C_n \otimes_h C_n$ and $C_n \otimes_h R_n$. By a formula in [**Pis2**], we have

$$(C_n \otimes_h C_n, C_n \otimes_h R_n)_\theta \cong C_n \otimes_h (C_n, R_n)_\theta \cong C_n(O\ell_n^p) \cong CB(R_n, O\ell_n^p),$$

where $O\ell_n^p$ is the standard operator space structure on ℓ_n^p from [**Pis1**]. Thus the operator space structure which we are putting on $L^p(M_n)$ is simply described as $C_n(O\ell_n^p)$. Equivalently, we are viewing the Schatten class operators on ℓ_n^2 as endowed with the $CB(R_n, O\ell_n^p)$ operator space structure. From this fact, as pointed out to us by Junge, the left adjointable multiplier algebra is easily computable. By fact (2.4) and the discussion at the end of Section 5.1, we have $\mathcal{A}_\ell(C_n(O\ell_n^p)) \cong M_n(\mathcal{A}_\ell(O\ell_n^p)) \cong M_n$, if $p < \infty$. From this and facts in the previous section, it is easy to see that the only right M-ideals in our interpolated space are of the form $J \otimes_h O\ell_n^p$, for a subspace J of C_n.

5.10. Infinite Matrices and Multipliers

We have the isometric identification $M_n(\mathcal{M}_\ell(X)) = \mathcal{M}_\ell(C_n(X))$ (see formula (2.4) from Section 2.1). For an infinite index set however, $M_I^w(\mathcal{M}_\ell(X))$ usually differs from $\mathcal{M}_\ell(C_I^w(X))$, even in the case that X is a unital C^*-algebra (see Appendix B for the relevant definitions). For dual spaces the situation is much better:

THEOREM 5.46. *Let X be a dual operator space, and let I, J be index sets. Then*

  (i) $\mathcal{A}_\ell(C_I^w(X)) \cong M_I^w(\mathcal{A}_\ell(X)) \cong \mathcal{A}_\ell(M_I^w(X))$ *as von Neumann algebras.*
  (ii) $\mathcal{A}_\ell(M_{I,J}^w(X)) \cong M_I^w(\mathcal{A}_\ell(X))$ *as von Neumann algebras.*
  (iii) $\mathcal{M}_\ell(M_{I,J}^w(X)) \cong M_I^w(\mathcal{M}_\ell(X))$ *as dual operator algebras (that is, via a completely isometric homomorphism which is a weak-* homeomorphism).*

Note that this result is only new if I, J are infinite (see [**B6**, Lemma 4.8] for the finite case). In the infinite case, the precise isomorphisms above are essentially the same as in the case that I, J are finite.

PROOF. (i) Suppose that $T \in \mathcal{A}_\ell(C_I^w(X))$. Fix $i, j \in I$, and let Q_i and U_{ij} be respectively the projection onto the ith entry, and the permutation of the i and j entries. We have $Q_i U_{ij} = U_{ij} Q_j$. It is easy to see that Q_i, U_{ij} are in $\mathcal{A}_\ell(C_I^w(X))$, and hence so is $S_{ij} = Q_j U_{ij} T Q_j$. Writing ϵ_i for the inclusion of X into $C_I^w(X)$ as the ith entry, we have $S_{ij}(\epsilon_j(X)) \subset \epsilon_j(X)$, and $S_{ij}^\star(\epsilon_j(X)) \subset \epsilon_j(X)$. Thus S_{ij} may be regarded as an element T_{ij} of $\mathcal{A}_\ell(X)$. Indeed $T_{ij} = \pi_i \circ T \circ \epsilon_j$, where $\pi_i : C_I^w(X) \to X$ is the ith coordinate function. We let $\theta(T) = [T_{ij}]$, and we will proceed to show that this is in $M_I^w(\mathcal{A}_\ell(X))$.

We introduce some notation: for a finite subset Δ of I, we define $P_\Delta = \sum_{i \in \Delta} Q_i$, the projection of $C_I^w(X)$ onto the space of columns 'supported on Δ'. For $T \in \mathcal{A}_\ell(C_I^w(X))$, let $T^\Delta = P_\Delta T P_\Delta$. Also, given any $I \times I$ matrix x, write x^Δ for the same matrix but with entries x_{ij} replaced by 0 if i or j is not in Δ. It is then easy to see that

$$\theta(T^\Delta) = \theta(T)^\Delta.$$

Let $A_\Delta \in M_\Delta(\mathcal{A}_\ell(X))$ be the 'Δ-submatrix' of $[T_{ij}]$. That is, A_Δ is the 'possibly nonzero part' of $\theta(T)^\Delta$. We identify $C_\Delta(X)$ with a subspace of $C_I^w(X)$, and notice that this subspace is invariant under T^Δ and its adjoint. We write $T^\Delta|_{C_\Delta(X)}$ for the restriction of T^Δ to this subspace; note that this restriction is in $\mathcal{A}_\ell(C_\Delta(X))$ by Lemma 5.5, and has the same norm there. Since $\mathcal{A}_\ell(C_n(X)) \cong M_n(\mathcal{A}_\ell(X))$ we have

$$\|A_\Delta\|_{M_\Delta(\mathcal{A}_\ell(X))} = \|T^\Delta|_{C_\Delta(X)}\|_{\mathcal{A}_\ell(C_\Delta(X))} = \|T^\Delta\|_{\mathcal{A}_\ell(C_I^w(X))} \le \|T\|_{\mathcal{A}_\ell(C_I^w(X))}.$$

From this we deduce that $\theta(T) = [T_{ij}] \in M_I^w(\mathcal{A}_\ell(X))$. Also we now see that θ is a unital linear contraction which is isometric on $P_\Delta \mathcal{A}_\ell(C_I^w(X))P_\Delta$.

Claim: θ is weak-* continuous. To see this take a bounded net (T_λ) converging to T with respect to the weak-* topology on $\mathcal{A}_\ell(C_I^w(X))$. By Theorem 2.3, this is equivalent to the fact that $T_\lambda(x) \to T(x)$ weak-* for all $x \in C_I^w(X)$. By the last remark in Appendix B, this is the same as saying that $\pi_i(T_\lambda(x)) \to \pi_i(T(x))$ weak-* for all $x \in C_I^w(X)$ and all $i \in I$. But then for any $x \in X$, $(\pi_i \circ T_\lambda \circ \epsilon_j)(x) \to (\pi_i \circ T \circ \epsilon_j)(x)$ in the weak-* topology on X. From Theorem 2.3 again, we conclude that $\pi_i \circ T_\lambda \circ \epsilon_j \to \pi_i \circ T \circ \epsilon_j$ with respect to the weak-* topology on $\mathcal{A}_\ell(X)$. Since the i-j entry of $\theta(T_\lambda)$ is $\pi_i \circ T_\lambda \circ \epsilon_j$, we see that $\theta(T_\lambda) \to \theta(T)$ weak-* in $M_I^w(\mathcal{A}_\ell(X))$ (Appendix B). By the Krein-Smulian Theorem A.1, θ is weak-* continuous.

To see that θ is an isometry we suppose that $\|T\| = 1$ but $\|\theta(T)\| \leq \alpha < 1$. Then

$$\|T^\Delta\| = \|\theta(T^\Delta)\| = \|\theta(T)^\Delta\| \leq \alpha < 1.$$

On the other hand, for any increasing net of projections $\{p_t\}$ in a von Neumann algebra M with supremum 1, it is well known that $p_t \to 1$ strongly on the underlying Hilbert space. Hence it is clear that the bounded net $p_t x p_t \to x$ in the weak-* topology for any $x \in M$. We conclude that $T^\Delta \to T$ weak-* in $\mathcal{A}_\ell(C_I^w(X))$, and it follows from the last centered equation that $\|T\| < 1$. This contradiction shows that θ is an isometry.

We may now deduce from the Krein-Smulian Theorem that θ has weak-* closed range. However, θ has weak-* dense range, by the last fact from Appendix B. Thus, θ is surjective. Finally, it is easy to see that θ restricted to $P_\Delta \mathcal{A}_\ell(C_I^w(X))P_\Delta$ is a *-homomorphism. By a routine weak-* approximation argument, we see that θ is a *-homomorphism. Thus, θ is a *-isomorphism of von Neumann algebras.

We have now proven that $\mathcal{A}_\ell(C_I^w(X)) \cong M_I^w(\mathcal{A}_\ell(X))$. We next claim that $\mathcal{A}_\ell(R_J^w(X)) \cong \mathcal{A}_\ell(X)$ as von Neumann algebras. One way to see this is to first observe that there is a weak-* continuous 1-1 *-homomorphism $\pi : \mathcal{A}_\ell(X) \to \mathcal{A}_\ell(R_J^w(X))$ (we leave this as an exercise). Conversely, if $T \in \mathcal{A}_\ell(R_J^w(X))_{sa}$ then T commutes with every $S \in \mathcal{M}_r(R_J^w(X))$. Hence T commutes with 'projection onto the ith entry', and so T is necessarily of the form $[x_i] \mapsto [T_i(x_i)]$ for maps T_i which are easily seen to be in $\mathcal{A}_\ell(X)$ (using Proposition 5.2). If Δ is as above then T restricts to a left multiplier of $R_\Delta(X)$, which forces $T_i = T_j$ for $i, j \in \Delta$. Thus $T_i = T_j$ for $i, j \in J$, and so π above is a surjection.

Finally,

$$\mathcal{A}_\ell(M_{I,J}^w(X)) \cong \mathcal{A}_\ell(R_J^w(C_I^w(X))) \cong \mathcal{A}_\ell(C_I^w(X)) \cong M_I^w(\mathcal{A}_\ell(X)),$$

establishing the final isomorphism in (i) and also (ii).

(iii) This is similar to the proof above, and we leave the details to the reader. The main obstacle encountered is that one may no longer appeal to Lemma 5.5 to see that the map from $P_\Delta \mathcal{M}_\ell(C_I^w(X))P_\Delta \to \mathcal{M}_\ell(C_\Delta(X))$ is isometric. We present one possible alternative. Let $T \in \mathcal{M}_\ell(C_\Delta(X))$, with $\|T\|_{\mathcal{M}_\ell(C_\Delta(X))} \leq 1$. By Proposition 1.6.4 of [Z1] (or the proof of [B6, Lemma 4.8]), there exists a complete isometry $\sigma : X \to B(H)$ and an $A \in M_\Delta(B(H))$, with $\|A\| \leq 1$, such that

$$\sigma_{\Delta,1}(Tx) = A\sigma_{\Delta,1}(x)$$

for all $x \in C_\Delta(X)$. Here $\sigma_{\Delta,1} : C_\Delta(X) \to C_\Delta(B(H))$ is the natural amplification of σ. Now let $\tilde{A} \in M_I^w(B(H))$ be the $I \times I$ matrix whose $\Delta \times \Delta$ submatrix equals A, and all of whose other entries equal 0. Clearly, $\tilde{A}\sigma_{I,1}(C_I^w(X)) \subset \sigma_{I,1}(C_I^w(X))$. Thus, there exists a $\tilde{T} \in \mathcal{M}_\ell(C_I^w(X))$ such that $\sigma_{I,1}(\tilde{T}x) = \tilde{A}\sigma_{I,1}(x)$ for all $x \in C_I^w(X)$. One has that $\|\tilde{T}\|_{\mathcal{M}_\ell(C_I^w(X))} \leq \|\tilde{A}\| = \|A\| \leq 1$. Also, $P_\Delta \tilde{T} P_\Delta|_{C_\Delta(X)} = T$. Since the choice of T was arbitrary, we conclude that the map $P_\Delta \mathcal{M}_\ell(C_I^w(X))P_\Delta \to \mathcal{M}_\ell(C_\Delta(X))$, which we knew was a one-to-one contraction, is also surjective, and its inverse is contractive. Hence it is isometric. As a final note, we observe that once one has the isometry in (iii), the complete isometry follows from the relation $M_n(\mathcal{M}_\ell(X)) \cong \mathcal{M}_\ell(M_n(X) \cong \mathcal{M}_\ell(C_n(X))$. $\qquad\square$

Remark. For a general operator space X there is at least a complete isometry $\mathcal{M}_\ell(C_I(X)) \to M_I^w(\mathcal{M}_\ell(X))$, and similarly for $\mathcal{A}_\ell(\cdot)$ in place of $\mathcal{M}_\ell(\cdot)$. To see the latter for example, we use the fact that $C_I(X)^{**} \cong C_I^w(X^{**})$ (see Appendix B). Thus by remarks below Proposition 5.14, $\mathcal{A}_\ell(C_I(X))$ may be regarded as a C^*-subalgebra of $\mathcal{A}_\ell(C_I^w(X^{**}))$. Consider the restriction ρ of the $*$-isomorphism $\mathcal{A}_\ell(C_I^w(X^{**})) \to M_I^w(\mathcal{A}_\ell(X^{**}))$ to the subalgebra $\mathcal{A}_\ell(C_I(X))$ of $\mathcal{A}_\ell(C_I^w(X^{**}))$. This is certainly a complete isometry. The second adjoint of the maps ϵ_i and π_i in the proof of Theorem 5.46 are the corresponding inclusion and projection maps between $C_I^w(X^{**})$ and its summands. Using this and the formula $T_{ij} = \pi_i \circ T \circ \epsilon_j$ from the proof of Theorem 5.46, it now easily follows that ρ maps into $M_I^w(\mathcal{A}_\ell(X))$.

The following proposition tells us that for a general operator space X, the diagonal elements of $M_I^w(\mathcal{M}_\ell(X))$ are contained isometrically in $\mathcal{M}_\ell(C_I^w(X))$. Although the result may be deduced from the last Theorem by going to the second dual, we give a simpler direct proof.

PROPOSITION 5.47. *Let X be an operator space, I be an index set, and $\{T_i : i \in I\}$ be a uniformly bounded collection of elements of $\mathcal{M}_\ell(X)$.*

(i) *The map*

$$\bigoplus_{i \in I} T_i : C_I^w(X) \to C_I^w(X) : \begin{bmatrix} \vdots \\ x_i \\ \vdots \end{bmatrix}_{i \in I} \mapsto \begin{bmatrix} \vdots \\ T_i x_i \\ \vdots \end{bmatrix}_{i \in I}$$

is an element of $\mathcal{M}_\ell(C_I^w(X))$, of multiplier norm $\sup_{i \in I} \|T_i\|_{\mathcal{M}_\ell(X)}$.

(ii) *$\bigoplus_{i \in I} T_i$ maps $C_I(X)$ into itself, and so restricts to a map in $\mathcal{M}_\ell(C_I(X))$. The restriction process is norm-preserving.*

(iii) *If each $T_i \in \mathcal{A}_\ell(X)$, then $\bigoplus_{i \in I} T_i \in \mathcal{A}_\ell(C_I^w(X))$, and $\left(\bigoplus_{i \in I} T_i\right)^* = \bigoplus_{i \in I} T_i^*$.*

PROOF. (i) Without loss of generality, we may assume that $\sup_{i \in I} \|T_i\|_{\mathcal{M}_\ell(X)} = 1$. It is not hard to see that $T = \bigoplus_{i \in I} T_i$ is a well-defined linear contraction on $C_I^w(X)$. Let $\sigma : X \to B(H)$ be a "Shilov embedding" of X (cf. [**BEZ**], §1). Then for each $i \in I$, there exists an $A_i \in B(H)$ such that (σ, A_i) is an implementing pair for T_i and $\|A_i\| = \|T_i\|_{\mathcal{M}_\ell(X)}$. It is easy to see that the I-fold 'column-amplification' of σ, namely the map $\sigma^I : C_I^w(X) \to C_I^w(B(H)) \cong B(H, H^{(I)})$ taking $[x_i]$ to $[\sigma(x_i)]$, is a complete isometry. Let $A = \bigoplus_{i \in I} A_i \in M_I^w(B(H)) \cong B(H^{(I)})$. Then (σ^I, A) is an implementing pair for T. Thus, $T \in \mathcal{M}_\ell(C_I^w(X))$ and $\|T\|_{\mathcal{M}_\ell(C_I^w(X))} \leq \|A\| = \sup_{i \in I} \|A_i\| = \sup_{i \in I} \|T_i\|_{\mathcal{M}_\ell(X)} = 1$. Now fix $i \in I$. Let X_i denote the closed linear subspace of $C_I^w(X)$ consisting of columns supported in the ith position. It is clear that T maps X_i into itself. Thus, $T|_{X_i} \in \mathcal{M}_\ell(X_i)$ and $\|T|_{X_i}\|_{\mathcal{M}_\ell(X_i)} \leq \|T\|_{\mathcal{M}_\ell(C_I^w(X))}$. But under the completely isometric identification $X_i = X$, $T|_{X_i}$ is identified with T_i. Thus, $\|T_i\|_{\mathcal{M}_\ell(X)} \leq \|T\|_{\mathcal{M}_\ell(C_I^w(X))}$. Since the choice of i was arbitrary, $\|T\|_{\mathcal{M}_\ell(C_I^w(X))} = 1$.

(ii) This statement is clear.

(iii) Suppose each $T_i \in \mathcal{A}_\ell(X)_{sa}$. Then each $A_i \in B(H)_{sa}$, which implies that the matrix A above is in $M_I^w(B(H))_{sa}$, which in turn implies that $T \in \mathcal{A}_\ell(C_I^w(X))_{sa}$. The result now follows. $\qquad \square$

COROLLARY 5.48. *Let $\{J_i : i \in I\}$ be a collection of right M-ideals (resp. right M-summands) of an operator space X. Then $\{[x_i] \in C_I(X) : x_i \in J_i\}$ is a right M-ideal of $C_I(X)$ (resp. $\{[x_i] \in C_I^w(X) : x_i \in J_i\}$ is a right M-summand of $C_I^w(X)$ and $\{[x_i] \in C_I(X) : x_i \in J_i\}$ is a right M-summand of $C_I(X)$).*

PROOF. First suppose that the J_i are right M-summands. If $\{P_i : i \in I\}$ is the matching collection of left M-projections, then by the previous proposition $P = \oplus_{i \in I} P_i$ is in $\mathcal{A}_\ell(C_I^w(X))_{sa}$, and it is easy to see that P is idempotent, and restricts to an idempotent $P' \in \mathcal{A}_\ell(C_I(X))_{sa}$. Thus P, P' are left M-projections, and the result is now clear.

Next suppose that the J_i are right M-ideals, so that $J_i^{\perp\perp}$ is a right M-summand of X^{**} for all i. By the first part, $K = \{[\eta_i] \in C_I^w(X^{**}) : \eta_i \in J_i^{\perp\perp}\}$ is a right M-summand of $C_I^w(X^{**})$. In particular it is weak-* closed. We will use the fact that $C_I(X)^{**} = C_I^w(X^{**})$. Let $J = \{[x_i] \in C_I(X) : x_i \in J_i\}$. Since $J \subset K$ we have $J^{\perp\perp} \subset K$. On the other hand, if $z \in K$, then the net of 'finitely supported submatrices' of z is a bounded net converging in the weak-* topology to z. Now any 'finitely supported' column in K is weak-* approximable by a bounded net of 'finitely supported' columns in J. Hence J is weak-* dense in K, so that $J^{\perp\perp} = K$. Thus J is a right M-ideal of $C_I(X)$. $\qquad \square$

Analogues of the last two results hold with $C_I^w(X)$ replaced by $M_{I,J}^w(X)$ for another index J, and with $C_I(X)$ replaced by $K_{I,J}(X)$. The proofs are almost identical, and are omitted. Thus for example we have:

COROLLARY 5.49. *Let J be a right M-ideal (resp. right M-summand) of an operator space X, and let I be an index set. Then $K_I(J)$ is a right M-ideal of $K_I(X)$ (resp. $M_I^w(J)$ is a right M-summand of $M_I^w(X)$, $M_I(J)$ is a right M-summand of $M_I(X)$, and $K_I(J)$ is a right M-summand of $K_I(X)$).*

5.11. Diagonal Sums

We write $\oplus_{i \in I}^\infty X_i$ for the usual 'ℓ^∞ direct sum' of operator spaces. We write $\oplus_i^0 X_i$ for the usual 'c_0 direct sum', that is, the closure in $\oplus_i^\infty X_i$ of the finitely supported tuples. Of course if I is finite, then these direct sums coincide. It is well known (see e.g. [**BLM**, Section 1.4]) that $(\oplus_i^1 X_i)^* = \oplus_i^\infty X_i^*$ completely isometrically where $\oplus_i^1 X_i$ is the 'ℓ^1 direct sum' of operator spaces, and that $(\oplus_i^0 X_i)^* = \oplus_i^1 X_i^*$ completely isometrically. Hence $(\oplus_i^0 X_i)^{**} = \oplus_i^\infty X_i^{**}$.

PROPOSITION 5.50. *If T_i are left multipliers (resp. left adjointable multipliers) on operator spaces X_i, with the multiplier norms of the T_i uniformly bounded above, then $\oplus_i^\infty T_i$ is a left multiplier (resp. left adjointable multiplier) on $\oplus_i^\infty X_i$. Moreover, the multiplier norm of $\oplus_i^\infty T_i$ is the supremum of the multiplier norms of the T_i. In the adjointable case, $(\oplus_i^\infty T_i)^\star = \oplus_i^\infty T_i^\star$.*

PROOF. This follows from an idea we have used repeatedly. Namely, suppose that (σ_i, S_i) is an implementing pair for T_i, for each $i \in I$. Let $S = \oplus_i^\infty S_i \in B(\oplus_i H_i)$, let $\sigma = \oplus_i^\infty \sigma_i : \oplus_i^\infty X_i \to B(\oplus_i H_i)$, and proceed as in the proof of Proposition 5.47.   □

COROLLARY 5.51. *Let $\{X_i : i \in I\}$ be a collection of operator spaces, and for each $i \in I$, suppose that J_i is a right M-ideal (resp. right M-summand) of X_i. Then $\{(x_i) \in \oplus_{i\in I}^0 X_i : x_i \in J_i\}$ is a right M-ideal of $\oplus_{i\in I}^0 X_i$ (resp. $\{(x_i) \in \oplus_{i\in I}^\infty X_i : x_i \in J_i\}$ is a right M-summand of $\oplus_{i\in I}^\infty X_i$ and $\{(x_i) \in \oplus_{i\in I}^0 X_i : x_i \in J_i\}$ is a right M-summand of $\oplus_{i\in I}^0 X_i$).*

PROOF. First suppose that the J_i are right M-summands. If $\{P_i : i \in I\}$ is the matching collection of left M-projections, then by the previous proposition $P = \oplus_{i\in I} P_i$ is in $\mathcal{A}_\ell(\oplus_{i\in I}^\infty X_i)_{sa}$, and it is easy to see that P is idempotent, and restricts to an idempotent $P' \in \mathcal{A}_\ell(\oplus_{i\in I}^0 X_i)_{sa}$. Thus P, P' are left M-projections, and the result is now clear.

Next suppose that the J_i are right M-ideals, so that the $J_i^{\perp\perp}$ are right M-summands of X^{**}. Then by the first part, $K = \{(\eta_i) \in \oplus_{i\in I}^\infty X_i^{**} : \eta_i \in J_i^{\perp\perp}\}$ is a right M-summand in $\oplus_{i\in I}^\infty X_i^{**}$. The rest of the argument is almost identical to that of Corollary 5.48, except that we use the fact that $(\oplus_{i\in I}^0 X_i)^{**} = \oplus_{i\in I}^\infty X_i^{**}$.   □

5.12. Mutually Orthogonal and Equivalent One-Sided M-Projections

LEMMA 5.52. *Let X be an operator space, $P_1, P_2, ..., P_n \in \mathcal{A}_\ell(X)$ be mutually orthogonal left M-projections, and $J_1, J_2, ..., J_n \subset X$ be the corresponding right M-summands. Then*

(i) *$P_1 + P_2 + ... + P_n$ is the left M-projection onto $J_1 + J_2 + ... + J_n$, and the complementary left M-projection is $P_1^\perp P_2^\perp ... P_n^\perp$. In particular, $J_1 + J_2 + ... + J_n$ is closed.*

(ii) *The map*

$$X \to C_{n+1}(X) : x \mapsto \begin{bmatrix} P_1 x \\ P_2 x \\ \vdots \\ P_n x \\ P_1^\perp P_2^\perp ... P_n^\perp x \end{bmatrix}$$

is a complete isometry.

(iii) *The map*

$$J_1 + J_2 + ... + J_n \to C_n(X) : x \mapsto \begin{bmatrix} P_1 x \\ P_2 x \\ \vdots \\ P_n x \end{bmatrix}$$

is a complete isometry.

PROOF. (i) Let $P = P_1 + P_2 + ... + P_n \in \mathcal{A}_\ell(X)$. It is easy to verify that P is a left M-projection (i.e. $P^2 = P$ and $P^\star = P$). Likewise, it is easy to see that the range of P is $J_1 + J_2 + ... + J_n$.

(ii) We proceed by induction on n. The case $n = 1$ follows immediately from Theorem 2.5. Now suppose that the statement is true for $n \geq 1$. That is, suppose that the map

$$X \to C_{n+1}(X) : x \mapsto \begin{bmatrix} P_1 x \\ P_2 x \\ \vdots \\ P_n x \\ P_1^\perp P_2^\perp ... P_n^\perp x \end{bmatrix}$$

is a complete isometry. Since, by Theorem 2.5 again, the map

$$\mu_{P_{n+1}}^c : X \to C_2(X) : x \mapsto \begin{bmatrix} P_{n+1} x \\ P_{n+1}^\perp x \end{bmatrix}$$

is a complete isometry, we conclude that the map

$$X \mapsto C_{2n+2}(X) : x \mapsto \begin{bmatrix} P_{n+1} P_1 x \\ P_{n+1}^\perp P_1 x \\ P_{n+1} P_2 x \\ P_{n+1}^\perp P_2 x \\ \vdots \\ P_{n+1} P_n x \\ P_{n+1}^\perp P_n x \\ P_{n+1} P_1^\perp P_2^\perp ... P_n^\perp x \\ P_{n+1}^\perp P_1^\perp P_2^\perp ... P_n^\perp x \end{bmatrix} = \begin{bmatrix} 0 \\ P_1 x \\ 0 \\ P_2 x \\ \vdots \\ 0 \\ P_n x \\ P_{n+1} x \\ P_1^\perp P_2^\perp ... P_n^\perp P_{n+1}^\perp x \end{bmatrix}$$

is a complete isometry. It follows that the statement is true for $n + 1$.

(iii) This is an immediate consequence of (ii). $\qquad\square$

COROLLARY 5.53. *Let X be an operator space and let $\{P_i : i \in I\} \subset \mathcal{A}_\ell(X)$ be a family of mutually orthogonal left M-projections.*

(i) *If $\sum_{i \in I} P_i = \mathrm{Id}_X$ in the point-norm topology, then the map*

$$X \to C_I(X) : x \mapsto \begin{bmatrix} \vdots \\ P_i x \\ \vdots \end{bmatrix}_{i \in I}$$

is a complete isometry.

(ii) *If X is a dual operator space and $\sum_{i \in I} P_i = \mathrm{Id}_X$ in the weak-* topology, then the map*

$$X \to C_I^w(X) : x \mapsto \begin{bmatrix} \vdots \\ P_i x \\ \vdots \end{bmatrix}_{i \in I}$$

is a weak- continuous complete isometry.*

PROOF. There is a quick proof of (ii) using 5.46. Instead we give another proof of (ii) which adapts easily to give (i).

(ii) Fix $x \in X$. Then $P_F x \to x$ weak-*, where for any finite subset F of I, $P_F = \sum_{i \in F} P_i$. It follows that $\|x\| \leq \sup_F \|P_F x\|$. Now suppose that $F = \{i_1, i_2, ..., i_n\} \subset I$. Then by Lemma 5.52,

$$\|P_F x\| = \left\| \begin{bmatrix} P_{i_1} x \\ P_{i_2} x \\ \vdots \\ P_{i_n} x \end{bmatrix} \right\| \leq \left\| \begin{bmatrix} \vdots \\ P_i x \\ \vdots \end{bmatrix}_{i \in I} \right\|.$$

Thus $\|x\| \leq \|[P_i(x)]\|$. On the other hand, for any $i_1, i_2, ..., i_n \in I$, one has by Lemma 5.52 that

$$\left\| \begin{bmatrix} P_{i_1} x \\ P_{i_2} x \\ \vdots \\ P_{i_n} x \end{bmatrix} \right\| \leq \left\| \begin{bmatrix} P_{i_1} x \\ P_{i_2} x \\ \vdots \\ P_{i_n} x \\ P_{i_1}^\perp P_{i_2}^\perp ... P_{i_n}^\perp x \end{bmatrix} \right\| = \|x\|,$$

which implies that $\|[P_i(x)]\| \leq \|x\|$. So the given map $X \to C_I^w(X)$ is an isometry. Since the n-fold amplification of this map may be identified with the map

$$M_n(X) \to C_I^w(M_n(X)) : x \mapsto \begin{bmatrix} \vdots \\ (P_i)_n x \\ \vdots \end{bmatrix}_{i \in I},$$

and since $\{(P_i)_n : i \in I\}$ is a family of mutually orthogonal left M-projections on the dual operator space $M_n(X)$ which add to the identity in the weak-* topology, the given map $X \to C_I^w(X)$ is in fact a complete isometry. That this map is weak-* continuous follows easily from the Krein-Smulian Theorem A.1 and the last remark in Appendix B, together with the fact that each P_i is weak-* continuous.

(i) This is similar to (ii). The only difference is that for fixed $x \in X$, we now have that $P_F x \to x$ in norm. This is a favorable difference—not only does it tell us that the given map is a complete isometry from X to $C_I^w(X)$, but also that the range lies in $C_I(X)$. $\qquad \square$

THEOREM 5.54. *Let X be an operator space and I be an index set.*

(i) *If I is finite, then X is completely isometric to $C_I(X_0)$ for some operator space X_0 if and only if there exists a family $\{P_i : i \in I\}$ of mutually orthogonal and equivalent left M-projections on X such that $\sum_{i \in I} P_i = \mathrm{Id}_X$.*

(ii) *If I is infinite, then X is completely isometric to $C_I(X_0)$ for some operator space X_0 if and only if there exists a family $\{P_i : i \in I\}$ of mutually orthogonal and equivalent left M-projections on X such that $\sum_{i \in I} P_i = \mathrm{Id}_X$ point-norm.*

(iii) *If I is infinite and X is a dual operator space, then X is completely isometric and weak-* homeomorphic to $C_I^w(X_0)$ for some dual operator space X_0 if and only if there exists a family $\{P_i : i \in I\}$ of mutually orthogonal and equivalent left M-projections on X such that $\sum_{i \in I} P_i = \mathrm{Id}_X$ weak-*.*

We have $\mathcal{A}_\ell(X_0) \cong P_i \mathcal{A}_\ell(X) P_i$ for each $i \in I$ and X_0 as above. If the equivalent conditions in (i) or (iii) hold, then we also have $\mathcal{A}_\ell(X) \cong M_I^w(\mathcal{A}_\ell(X_0))$.

PROOF. We will only prove (iii) and the statement following it. The proof of (ii) is similar, and (i) follows immediately from (ii).

($\Rightarrow$) Suppose $X = C_I^w(X_0)$ for some dual operator space X_0. For each $i \in I$, let $P_i : C_I^w(X_0) \to C_I^w(X_0)$ be the projection onto the ith coordinate. Then P_i is a left M-projection. Obviously, the P_i's are mutually orthogonal. We claim that they add to the identity in the weak-* topology. To see this, fix $x \in C_I^w(X)$ and $j \in I$. If $F \subset I$ is a finite set containing j and $P_F = \sum_{i \in F} P_i$, then $(P_F x)_j = x_j$. Consequently, $(P_F x)_j \to x_j$ weak-*. Since the choice of j was arbitrary, $P_F x \to x$ weak-* (see Appendix B). Since the choice of x was arbitrary, the claim is proven. To see that the P_i's are mutually equivalent, fix $i, j \in I$ with $i \neq j$. Let $U_{ij} : C_I^w(X_0) \to C_I^w(X_0)$ be the map which swaps the i and j coordinates. Then U_{ij} is a unitary element of $\mathcal{A}_\ell(C_I^w(X_0))$. Now define $V = U_{ij} P_i$. Then $V \in \mathcal{A}_\ell(C_I^w(X_0))$, $V^\star V = P_i$, and $V V^\star = P_j$. Thus, $P_i \sim P_j$.

($\Leftarrow$) Suppose there exists a family $\{P_i : i \in I\}$ of mutually orthogonal and equivalent left M-projections on X such that $\sum_{i \in I} P_i = \mathrm{Id}_X$ in the weak-* topology. Fix $i \in I$. For each $j \in I$, let $V_j \in \mathcal{A}_\ell(X)$ be such that $V_j^* V_j = P_j$ and $V_j V_j^* = P_i$. For the sake of definiteness, let $V_i = P_i$. By Proposition 5.47, $V \equiv \bigoplus_{j \in I} V_j$ is a partial isometry in $\mathcal{A}_\ell(C_I^w(X))$ with range $C_I^w(P_i(X))$. Note that V is weak-* continuous by a fact listed after Theorem 2.3. Also, V is completely isometric on the range of the map

$$\theta : X \to C_I^w(X) : x \mapsto \begin{bmatrix} \vdots \\ P_j x \\ \vdots \end{bmatrix}_{j \in I}.$$

Since, by Corollary 5.53, θ itself is completely isometric and weak-* continuous, the composition $R = V \circ \theta$ is a completely isometric weak-* continuous isomorphism of X with $C_I^w(P_i(X))$. However $P_i(X)$ is a right M-summand of a dual operator space. Thus by a basic fact mentioned after Theorem 2.5, we have that $P_i(X)$ is weak-* closed, and hence is a dual operator space. By the Krein-Smulian theorem A.1 we see that R is a homeomorphism for the weak-* topologies.

Assuming that the equivalent conditions in (iii) hold, the fact that $\mathcal{A}_\ell(X) \cong M_I^w(\mathcal{A}_\ell(X_0))$ follows from Theorem 5.46. To express $\mathcal{A}_\ell(X_0)$ in terms of $\mathcal{A}_\ell(X)$, note that by Lemma 5.5 we need only show that the 'inclusion' $P_i \mathcal{A}_\ell(X) P_i \to \mathcal{A}_\ell(X_0)$ is surjective. Since $X \cong C_I^w(X_0)$, one can check that it suffices to show that the 'inclusion' $Q_i \mathcal{A}_\ell(C_I^w(X_0)) Q_i \to \mathcal{A}_\ell(X_0)$ is surjective, where Q_i is as in the proof of Theorem 5.46. However this is a simple consequence of Proposition 5.47 (iii) for example. $\qquad\square$

Remark. If, in addition to the equivalent conditions in each part of the last theorem, we assume that each P_i is equivalent to the identity map on X, then one can deduce that X is 'column stable', in the sense that $X \cong C_I(X)$ completely isometrically in (i) or (ii), and that $X \cong C_I^w(X)$ completely isometrically via a weak-* homeomorphism in (iii). The argument for this is quite straightforward: Recall that the X_0 mentioned in (i)–(iii) is $P_i(X)$ for one fixed $i \in I$. Since P_i is equivalent to Id_X, the desired result follows from facts towards the end of Section 3.3.

COROLLARY 5.55. Let X be a dual operator space. If X has no complete right M-summand which is completely isometric to $C_2(X_0)$ for an operator space X_0, then all left M-projections on X commute. In particular, if X does not contain C_2 completely isometrically, then all left M-projections on X commute.

PROOF. If $\mathcal{A}_\ell(X)$ is noncommutative, then by the last remark in Section 3.3, there exist two equivalent mutually orthogonal left M-projections P, Q. Suppose that $VV^\star = P, V^\star V = Q$. Let $R = P + Q$, another left M-projection, and consider the von Neumann algebra $R \mathcal{A}_\ell(X) R$, which may be viewed as a von Neumann subalgebra of $\mathcal{A}_\ell(R(X))$, using Lemma 5.5. Then $VR = VQR = VQ = V$ and $RV = RPV = PV = V$. Thus $V = RVR \in R\mathcal{A}_\ell(X)R$, and hence $P \sim Q$ in $\mathcal{A}_\ell(R(X))$. Appealing to Theorem 5.54 (i) we have $R(X) \cong C_2(X_0)$. □

Remark. The converse is false. Indeed $X = \ell^\infty \otimes_{\mathrm{h}} C_2$ is a dual operator space which contains C_2 completely isometrically, but $\mathcal{A}_\ell(X) = \ell^\infty$ (see the proof of Corollary 5.43), which is commutative.

5.13. Multiple $\oplus_{\mathrm{rM}}$-sums

Suppose that $\{P_i : i \in I\}$ is a collection of mutually orthogonal left M-projections on an operator space (resp. dual operator space) X which add up to Id_X in the point-norm (resp. weak-*) topology. Then we say that X is an '$\oplus_{\mathrm{rM}}$-sum' of the J_i, and write $X = \oplus_{\mathrm{rM}} J_i$ (resp. $X = \bar{\oplus}_{\mathrm{rM}} J_i$). We shall need the 'weak-*' case in Section 6.1. By Corollary 5.53 it follows that if $X = \oplus_{\mathrm{rM}} J_i$ (resp. $X = \bar{\oplus}_{\mathrm{rM}} J_i$), then the map $x \mapsto [P_i(x)]$ from X into $C_I(X)$ (resp. $C_I^w(X)$) is a complete isometry (resp. weak-* continuous complete isometry). We now establish the converse to this. For simplicity we will first restrict our attention to the case that I is finite; and then we simply write $X = J_1 \oplus_{\mathrm{rM}} J_2 \oplus_{\mathrm{rM}} \cdots \oplus_{\mathrm{rM}} J_n$.

PROPOSITION 5.56. Let X be an operator space, and suppose that $P_1, P_2, ..., P_n$ are linear projections on X. Then the following are equivalent:

  (i) $P_1, P_2, ..., P_n$ are left M-projections with sum Id_X.
  (ii) $P_i P_j = 0$ for all $i \neq j$, and the map $x \mapsto [P_i(x)]$ from X into $C_n(X)$ is a complete isometry.
  (iii) $P_i(X) \wedge P_j(X)$ if $i \neq j$, and $\sum_i P_i = \mathrm{Id}_X$.
  (iv) There exists a complete isometry $\sigma : X \to B(H)$ and orthogonal projections $E_1, E_2, ..., E_n \in B(H)$ such that (σ, E_i) is an implementing pair for P_i, $1 \leq i \leq n$, and $E_1 + E_2 + ... + E_n = I$.

PROOF. (iv) $\Rightarrow$ (iii) The condition implies that the E_i are mutually orthogonal. Thus $\sigma(P_i(x))^* \sigma(P_j(y)) = 0$ if $i \neq j$. Note also that $\sigma(\sum_i P_i(x)) = \sum_i E_i \sigma(x) = \sigma(x)$ for all $x \in X$.

  (iii) $\Rightarrow$ (i) If $\langle \cdot, \cdot \rangle$ is the 'left Shilov inner product' (see Section 3.4), then $\langle P_i(x), P_j(y) \rangle = 0$ if $i \neq j$. Hence

$$\langle P_i(x), y \rangle = \langle P_i(x), \sum_j P_j(y) \rangle = \langle P_i(x), P_i(y) \rangle.$$

Likewise,

$$\langle x, P_i(y) \rangle = \langle P_i(x), P_i(y) \rangle.$$

Thus $P_i \in \mathcal{A}_\ell(X)_{sa}$. This gives (i).

(i) $\Rightarrow$ (ii) The assumption gives us that $P_i P_j = 0$ if $i \neq j$. Now apply Lemma 5.52.

(ii) $\Rightarrow$ (iv) Let ν^c be the map in (ii). Suppose that $\rho : X \to B(H)$ is any complete isometry, and define $\sigma : X \to M_n(B(H))$ to be the map taking x to the matrix with first column $[\rho(P_i(x))]$ and other columns 0. Let E_i be the projection in $M_n(B(H))$ which is 0 except for an I in the i-i entry. Then $E_i \sigma(x) = \sigma(P_i(x))$. Clearly $\sum_i E_i = I$. $\qquad \square$

Remark. A variant of this result holds with an almost identical proof for a collection $\{P_i : i \in I\}$ of linear projections on X. In this case one must interpret the convergence of the sums in (i) and (iii) in the 'point-norm' topology, and in (iv) one insists that $\sum_i E_i \sigma(x)$ converges in norm to $\sigma(x)$ for all x.

We now turn to the 'weak-* case', which we will need in Section 6.1. One may view the sum $\bar{\oplus}_{\mathrm{rM}} J_i$ as a generalization in some sense of the 'ultraweak direct sum' of self-dual C^*-modules over a W^*-algebra [**Pas**]. We have the following:

PROPOSITION 5.57. Let X be a dual operator space, and suppose that $\{P_i : i \in I\}$ is a collection of linear projections on X. Then the following are equivalent:

(i) The P_i are left M-projections with $\sum_i P_i = \mathrm{Id}_X$ in the weak-* topology of $\mathcal{A}_\ell(X)$.

(ii) $P_i P_j = 0$ if $i \neq j$, and the map $x \mapsto [P_i(x)]$ from X into $C_I^w(X)$ is a weak-* continuous complete isometry.

(iii) $\sum_i P_i = \mathrm{Id}_X$ in the point-weak-* topology, and there is a weak-* continuous complete isometry σ from X into a von Neumann algebra with $\sigma(P_i(X))^* \sigma(P_j(X)) = 0$ whenever $i \neq j$.

(iv) There exists a weak-* continuous complete isometry $\sigma : X \to B(H)$ and orthogonal projections $E_i \in B(H)$, $i \in I$, such that (σ, E_i) is an implementing pair for P_i, $i \in I$, and $\sum_i E_i = I$ (in the weak operator topology).

PROOF. The only change to the proof that (iv) implies (iii) is an appeal to the Krein-Smulian theorem A.1 to see that σ is a homeomorphism for the weak-* topologies, from which it follows that $\sum_i P_i(x) = x$ in the weak-* topology for each $x \in X$. In the proof that (iii) implies (i) we replace the inner product by $\sigma(\cdot)^* \sigma(\cdot)$. Also note that if $P_i \in \mathcal{A}_\ell(X)$, then saying that $\sum_i P_i = \mathrm{Id}_X$ in the point-weak-* topology is the same as saying that $\sum_i P_i = \mathrm{Id}_X$ in the weak-* topology of $\mathcal{A}_\ell(X)$. In the proof that (ii) implies (iv) we take ρ to also be weak-* continuous. $\qquad \square$

The $\oplus_{\mathrm{rM}}$-sum is *associative*. To see this, suppose for specificity that X is an operator space with $X = Y \oplus_{\mathrm{rM}} W \oplus_{\mathrm{rM}} Z$, for subspaces Y, W, Z of X. Then clearly Y, W, Z are all right M-summands of X. Also $Y + W, Y + Z, W + Z$, are all right M-summands of X. Indeed $Y + W$ is the complementary right M-summand of Z in X, for example. These statements are all easiest seen using obvious properties of projections in C^*-algebras.

Also note that if Y, J are right M-summands of X, with $Y \subset J$, then Y is a right M-summand of J (by Proposition 5.3), and if W is the complementary right M-summand of Y in J, then it is easy to see that W is a right M-summand of X, and we have $X = Y \oplus_{\mathrm{rM}} W \oplus_{\mathrm{rM}} Z$, where Z is the complementary right M-summand of J in X.

Remark. It is easy to check, by a slight variation of the argument in Section 5.6, that (finite or 'point-norm') $\oplus_{r M}$-sums in an ambient operator space X distribute over a Haagerup or spatial tensor product (for example) with a fixed operator space V.

CHAPTER 6

One-Sided Type Decompositions and Morita Equivalence

6.1. One-Sided Type Decompositions

In this section X is a dual operator space, so that $\mathcal{A}_\ell(X)$ is a W^*-algebra and therefore has a type decomposition (see any book on von Neumann algebras). We say that a right M-summand J of X is finite, infinite, or properly infinite, according to whether the associated left M-projection P has this property as a projection in $\mathcal{A}_\ell(X)$ (see von Neumann algebra texts for the definitions). Strictly speaking one should perhaps say 'left finite', 'left infinite', or 'left properly infinite', but we suppress the 'left' in the following to avoid excessive verbiage. We will use the fact that if Id_X is not finite, then it is a unique sum of a central finite projection P_f (possibly 0) and a central properly infinite projection P_{pi}. Thus if X is not finite, we can write canonically $X = X_f \oplus_{\mathrm{rM}} X_{pi}$, where $X_f = P_f(X), X_{pi} = P_{pi}(X)$. Similarly, we may decompose the identity Id_X as a sum of central projections corresponding to the type decomposition of $\mathcal{A}_\ell(X)$. We say for example, if P is the central projection in this decomposition such that $\mathcal{A}_\ell(X)P$ is type II_1, that the summand $P(X)$ is the *type II_1 summand* of X. Thus we can write uniquely X as a $\oplus_{\mathrm{rM}}$-direct sum (see section 5.13) of a finite type I summand (or (0)), a purely infinite type I summand (or (0)), a type II_1 summand (or (0)), a type II_∞ summand (or (0)), and a type III summand (or (0)). The type I summand can be further written as a $\bar{\oplus}_{rM}$-sum of type I_n summands of X, for appropriate cardinals n. We say for example that X is of (left) type II_1 if $\mathcal{A}_\ell(X)$ is a type II_1 W^*-algebra.

In this section we present a few simple consequences of this type decomposition.

In regard to (iii) and (iv) of the next theorem, we recall that to say 'X contains a completely contractively complemented copy of Y' means that there is a completely contractive projection of X onto a subspace of X which is completely isometrically isomorphic to Y.

THEOREM 6.1. *Let X be a dual operator space.*

(i) *If X is of type I_n for a cardinal n, then X is completely isometric and weak-* homeomorphic to $C_n^w(X_0)$, for a dual operator space X_0 which has the property that all its left M-projections commute.*

(ii) *If X contains a nontrivial type I_n summand in its type decomposition (n possibly infinite), then this summand is completely isometric and weak-* homeomorphic to $C_n^w(X_0)$, for some dual operator space X_0. Similarly for a nontrivial type II_1 summand, except that now n can be taken to be any positive integer.*

(iii) *X is properly infinite if and only if X is 'column stable' in the sense that $X \cong C_\infty^w(X)$ completely isometrically and weak-* homeomorphically.*

Indeed if X is not finite, then for the properly infinite summand W of X we have $W \cong C_\infty^w(W)$. In this case, X contains a completely contractively complemented infinite dimensional Hilbert column space.

(iv) If X is not of type I, then X contains completely contractively complemented copies of Hilbert column space of arbitrarily large finite dimension. Hence, any dual operator space X for which there is a finite n such that X does not contain a copy of C_n is of type I.

(v) If there are no left M-projections Q on X with $Q \mathcal{A}_\ell(X) Q$ abelian (i.e. if X has no type I summand), then for every finite integer n we have that X is completely isometric and weak-$*$ homeomorphic to $C_n(X_0)$, for some dual operator space X_0.

The X_0 in (i), (ii), and (v) may be taken to be a right M-summand of X.

PROOF. We will treat (ii) in detail; and it may be used as a model for any omitted details in other parts.

(ii) Suppose that P is the central projection in $\mathcal{A}_\ell(X)$ corresponding to the type II$_1$ summand, and set $W = P(X)$. By Lemma 5.5, we may regard $\mathcal{A}_\ell(X)P$ as a unital C^*-subalgebra of $\mathcal{A}_\ell(W)$. By [**KR2**, Lemma 6.5.6], for any finite n, there are n mutually orthogonal and Murray-von Neumann equivalent projections $P_i \in \mathcal{A}_\ell(X)P$ adding up to P. It is clear that the restrictions of these projections give n mutually orthogonal and Murray-von Neumann equivalent projections in $\mathcal{A}_\ell(W)$ adding up to Id$_W$. Thus by Theorem 5.54 and the remark after it, we have that W is completely isometric and weak-$*$ homeomorphic to $C_n(X_0)$, for a dual operator space X_0. In fact, X_0 may be taken to be the range of $P_i|_W$ (any i). But this is the same as the range of P_i, which is a right M-summand of X.

Similarly, let Y_n be the type I$_n$ piece of X, and consider the central left M-projection Q onto Y_n. Then Q is the sum of n equivalent mutually orthogonal (abelian) projections in $\mathcal{A}_\ell(Y_n)Q$. This implies, as in the last paragraph, that $Y_n \cong C_n^w(X_0)$ completely isometrically and weak-$*$ homeomorphically, for a subspace X_0 of Y_n which is a right M-summand of X.

(iii) If X is properly infinite then by [**Sak2**, Proposition 2.2.4], Id$_X = \sum_i P_i$ for a countably infinite family of mutually orthogonal left M-projections, each equivalent to Id$_X$. Now apply Theorem 5.54 (iii), and the remark after that result. If X is 'column stable' then it is easy to find a sequence of mutually orthogonal projections in $\mathcal{A}_\ell(X)$, each Murray-von Neumann equivalent to Id_X, and adding up to Id_X. This is easily seen to imply that X is properly infinite.

The second part of (iii) follows from the first part; we leave the details as an exercise. Recall that a Hilbert column space is *injective* (being a 'corner' of $B(H)$), so that it is automatically completely contractively complemented.

(iv) This follows from (ii) and (iii).

(v) Follows from [**KR2**, Lemma 6.5.6] and Theorem 5.54 (i), as in the first paragraph of the proof of (ii).

(i) By (ii) we have $X \cong C_n^w(X_0)$. By the proof of (ii) and by the last assertion of Theorem 5.54, we see that $\mathcal{A}_\ell(X_0) = P_i \mathcal{A}_\ell(X) P_i$ is abelian (recall that the P_i here are abelian projections). $\qquad\square$

One may easily deduce from the last Theorem the fact mentioned earlier that if X does not contain a copy of C_2, then all left M-projections on X commute.

On the other hand, the opposite extreme to having all left M-projections commute would be the following:

COROLLARY 6.2. *Suppose that X is a dual operator space with the property that there are no nontrivial left M-projections on X which commute with all other left M-projections. Then one and only one of the following cases occurs:*

(i) $X \cong C_\infty^w(X)$ *completely isometrically and weak-* homeomorphically.*

(ii) X *is of finite type, and for all finite n there exists a dual operator space X_0 such that we may write $X \cong C_n(X_0)$ completely isometrically and weak-* homeomorphically.*

(iii) $X \cong C_n(Z)$ *completely isometrically and weak-* homeomorphically for a finite integer n, and a dual operator space Z which has no nontrivial left M-projections at all.*

The X_0 and Z here may be taken to be right M-summands of X.

PROOF. The condition asserts that $\mathcal{A}_\ell(X)$ is a factor. If $\mathcal{A}_\ell(X)$ is infinite, then it is properly infinite and we may appeal to Theorem 6.1 (iii). If $\mathcal{A}_\ell(X)$ is type II_1 (and thus finite), then we appeal to Theorem 6.1 (ii). If X is finite type I, then $\mathcal{A}_\ell(X) \cong M_n$ for a finite n, and $X \cong C_n(X_0)$ as in Theorem 5.54. By the last assertion of Theorem 5.54 we have $M_n(\mathcal{A}_\ell(X_0)) \cong M_n$. Thus $\mathcal{A}_\ell(X_0)$ is one-dimensional.

Finally, since $\mathcal{A}_\ell(C_n(X_0)) \cong M_n(\mathcal{A}_\ell(X_0))$, we see that cases (ii) and (iii) are mutually exclusive. So is case (i), that being the properly infinite case. $\square$

Remark: We have not explored the connection between our type decomposition and the weak-* TRO type decomposition that can be found in the recent paper [**Rua**] for example. However we imagine that this connection is largely a formal one; we certainly do not have access in our general framework to the kinds of properties that weak-* TRO's possess.

EXAMPLE 6.3. We give an (dramatic) illustration of the obstacle that arises in using such type decompositions to study dual operator spaces. Namely, one may have a dual operator space X with type decomposition $X = X_1 \oplus_{\mathrm{rM}} X_2$ for example, where X_1 is a type I_1 summand and X_2 is a type I_m summand ($m \neq 1$), but as an operator space in its own right, X_1 is of type II_1 say. This shows that a type I space can have a right M-summand which relative to itself may be of quite different type. Of course this bad behavior cannot occur if the summands are hereditary in the sense of Definition 5.7 (see also Proposition 5.6).

To construct such an example, let $\mathcal{R}$ be the hyperfinite type II_1 factor, and let τ be its trace. We fix an integer $m > 1$ and identify $M_m \cong B(\mathbb{C}^m)$. Let X be the subspace of $B(H \oplus \mathbb{C}^m \oplus \mathbb{C}^m)$ consisting of matrices of the form

$$\begin{bmatrix} a & 0 & 0 \\ 0 & \tau(a)I_m & 0 \\ 0 & B & 0 \end{bmatrix},$$

for $a \in \mathcal{R}, B \in M_m$. We will use basic facts about triples and the triple envelope from [**Ham**]. The triple subsystem Z generated by X in $B(H \oplus \mathbb{C}^m \oplus \mathbb{C}^m)$ is easily seen to be the subspace consisting of matrices of the form

$$\begin{bmatrix} a & 0 & 0 \\ 0 & C & 0 \\ 0 & B & 0 \end{bmatrix},$$

for $a \in \mathcal{R}$ and $B, C \in M_m$. It is easy to show that Z has only two nontrivial 'triple ideals' (using the well known facts that $\mathcal{R}$ and M_m are simple, and that triple ideals in Z are in a 1-1 correspondence with two-sided ideals in the C^*-algebra Z^*Z), namely the ones isomorphic to $\mathcal{R} \oplus 0$ and $0 \oplus M_{2m,m}$. From this it follows that Z is the triple envelope of X. Thus the projections in $\mathcal{A}_\ell(X)$ are easily characterized: they are exactly all the projections of the form $\lambda 1_\mathcal{R} \oplus \lambda I_m \oplus P$, for $\lambda = 0$ or 1 in $\mathbb{C}$, and projections $P \in M_m$. Since $\mathcal{A}_\ell(X)$ is the span of these projections we see that $\mathcal{A}_\ell(X) \cong \mathbb{C} \oplus M_m$. Thus in the type decomposition of X we only have a type I_1 and a type I_m summand. However if P is the projection corresponding to the type I_1 summand, then we see that $P(X)$ is type II_1 relative to itself.

The calculations in the last Example work with $\mathcal{R}$ replaced by any simple C^*-algebra satisfying a reasonable extra condition.

Finally we remark on what it means to say that an operator space X is (left) type II_1, for example. At this point the main significance we see is as follows. Hitherto in operator space theory the only 'column sums' considered are of the form $C_I(X)$ or $C_I^w(X)$. We believe that a 'left type II_1 operator space', for example, is a generalized kind of 'column sum'. Dual operator spaces of types I–III simultaneously generalize von Neumann algebras of these types, and also 'operator space column sums'. Such generalized column and row sums should play a further role in operator space theory.

6.2. Morita Equivalence

In this section we highlight a connection between our work and the very well understood Morita equivalence of von Neumann algebras [**Rie**]. One of our key tools will be Theorem 5.46.

DEFINITION 6.4. Two dual operator spaces X and Y are said to be *weakly column Morita equivalent* if there is a cardinal I such that $C_I^w(X) \cong C_I^w(Y)$ via a weak-* continuous complete isometry. There is a similar definition for (weak) *row Morita equivalence*. We say that X and Y are *weakly Morita equivalent* if there are cardinals I, J such that $M_{I,J}^w(X) \cong M_{I,J}^w(Y)$ via a weak-* continuous complete isometry.

Clearly weak column or row Morita equivalence implies weak Morita equivalence. We recall the Morita equivalence of von Neumann algebras investigated in [**Rie**]; and as some other authors do, we shall call this notion *weak Morita equivalence* of von Neumann algebras.

PROPOSITION 6.5. If dual operator spaces X and Y are weakly Morita equivalent, then the von Neumann algebras $\mathcal{A}_\ell(X)$ and $\mathcal{A}_\ell(Y)$ are weakly Morita equivalent as von Neumann algebras (i.e. in the sense of [**Rie**]). Thus $M_I^w(\mathcal{A}_\ell(X)) \cong M_I^w(\mathcal{A}_\ell(Y))$ *-isomorphically as von Neumann algebras, for some index set I.

PROOF. It is well known that von Neumann algebras M and N are Morita equivalent in the sense of [**Rie**] if and only if $M \bar{\otimes} B(H) \cong N \bar{\otimes} B(H)$ for some Hilbert space H (see e.g. [**BLM**, Chapter 8]). Thus the result follows from Theorem 5.46. $\square$

Remarks. 1) It follows that von Neumann algebras are weakly Morita equivalent in the sense of Definition 6.5 if and only if they are weakly Morita equivalent as von Neumann algebras (i.e. in the sense of [**Rie**]).

2) By taking second duals it follows that if operator spaces X and Y are *column stably isomorphic* (that is, if $C_I(X) \cong C_I(Y)$ completely isometrically), then X^{**} and Y^{**} are weakly column Morita equivalent. One may deduce from this, for example, that in this case the right M-ideals of X corresponding to projections in the center of $\mathcal{A}_\ell(X^{**})$ are in a bijective correspondence with the corresponding set of right M-ideals of Y. We omit the details.

Since we are not aware of the following result in the literature, we include a proof of it supplied by David Sherman.

LEMMA 6.6. *Let M, N be von Neumann algebras with orthogonal projections $p \in M, q \in N$ such that $M \cong qNq$ and $N \cong pMp$ as von Neumann algebras. Then M and N are weakly Morita equivalent. That is, $M \bar{\otimes} B(H) \cong N \bar{\otimes} B(H)$ for some Hilbert space H.*

PROOF. Let e, f be the central supports of p, q. It is well known that pMp is weakly Morita equivalent to eM. Thus $N \bar{\otimes} B(H) \cong (e \otimes 1)(M \bar{\otimes} B(H))$, for some Hilbert space H. A similar relation holds for $M \bar{\otimes} B(K)$. We may assume without loss of generality, by replacing H and K by Hilbert spaces of larger dimension, that $H = K$. Replacing M and N by their tensor products with $B(H)$, we see that it suffices to prove our Lemma in the case that p, q are central projections. So henceforth suppose that p, q are central, and we will in fact deduce that in this case $M \cong N$ as von Neumann algebras.

Let $f : M \to qN, g : N \to pM$ be the $*$-isomorphisms. We claim that f maps the center of M into the center of N. For if x is in the center of M and $y \in N$ then

$$f(x)y = f(x)qy = f(x)f(z) = f(xz) = f(zx) = yf(x),$$

for some $z \in M$. Similarly g maps between centers. With some work, one can check that the family $\{(fg)^k(1_N - q), (fg)^k f(1_M - p) : k = 0, 1, 2, ...\}$ consists of mutually orthogonal central projections. Let $d = \sum_k (fg)^k(1_N - q)$ and $e = \sum_k (fg)^k f(1_M - p)$, then $de = 0$. Also, $\{(fg)^k(1_N) : k = 0, 1, 2, \cdots\}$ is a decreasing sequence of projections; let r be its infimum or weak limit. Since $\sum_{k=0}^n (fg)^k(1_N - q) + \sum_{k=0}^n (fg)^k f(1_M - p) = 1_N - (fg)^{n+1}(1_N)$, it follows that $d + e + r = 1_N$. Thus we may centrally decompose N as a direct sum of three pieces Nd, Ne, Nr. Similarly M is a direct sum of three matching pieces Ma, Mb, Mc. It is easy to see that g restricts to an isomorphism of Nd with Mb, and that f^{-1} restricts to an isomorphism of Ne with Ma. To complete the argument we will show that g restricts to an isomorphism of Nr with Mc.

To that end, note that $x \in Nr$ if and only if $x = (fg)^k(1_N)x$ for all $k = 0, 1, 2, ...$ A similar statement holds for Mc. For such x, $g(x) = (gf)^k(g(1_N))g(x)$. Hence $(gf)^k(1_M)g(x) = g(x)$ for all k, so that $g(x) \in Mc$. To see that $Mc \subset g(Nr)$, take $y \in Mc$. Thus $(gf)^k(1_M)y = y$ for all k. Let $x = g^{-1}(y)$, then

$$g(x) = y = (gf)^k(1_M)y = g((fg)^{k-1}f(1_M)x).$$

Thus $x = (fg)^{k-1}f(1_M)x$ for all $k \in \mathbb{N}$, so that $x \in Nr$.   □

The following was found during a conversation with David Sherman:

PROPOSITION 6.7. Suppose that X is a dual operator space and that Y is a hereditary (see Definition 5.7) right M-summand of X (or more generally of $C_I^w(X)$ for a cardinal I). Then $\mathcal{A}_\ell(X)$ is weakly Morita equivalent to $\mathcal{A}_\ell(Y)$ if either of the following two conditions hold:

 (i) The central support of the left M-projection onto Y is Id_X (or more generally $\mathrm{Id}_{C_I^w(X)}$).
 (ii) There is a hereditary right M-summand of $C_J^w(Y)$ which is completely isometric and weak-* homeomorphic to X.

PROOF. The hypothesis implies that $\mathcal{A}_\ell(Y) \cong PMP$ as von Neumann algebras, where P is the left M-projection onto Y, and $M = \mathcal{A}_\ell(C_I^w(X))$. If, in addition, condition (i) holds, then P is a so-called *full projection*, and it is well known that in this case PMP is weakly Morita equivalent to M. On the other hand, $\mathcal{A}_\ell(C_I^w(X)) \cong M_I^w(\mathcal{A}_\ell(X))$, which is weakly Morita equivalent to $\mathcal{A}_\ell(X)$.

Under assumption (ii) we also have $\mathcal{A}_\ell(X) \cong Q\,\mathcal{A}_\ell(C_J^w(Y))Q$, and again we recall that $\mathcal{A}_\ell(C_J^w(Y)) \cong M_J^w(\mathcal{A}_\ell(Y)) \cong \mathcal{A}_\ell(Y)\bar{\otimes}B(H)$, for some Hilbert space H. As in the first lines of the proof of Lemma 6.6, this implies that $\mathcal{A}_\ell(X)\bar{\otimes}B(\tilde{H})$ is *-isomorphic to $e(\mathcal{A}_\ell(Y)\bar{\otimes}B(\tilde{H}))$ for a central projection $e \in \mathcal{A}_\ell(Y)\bar{\otimes}B(\tilde{H})$. A similar relation holds for $\mathcal{A}_\ell(Y)\bar{\otimes}B(K)$ for some Hilbert space K. We may assume that $\tilde{H} = K$, as in the last proof, and then the result follows from Lemma 6.6. $\square$

Remark. Thus if X and Y are two dual operator spaces with Y completely isometric to a hereditary right M-summand of X, and X completely isometric to a hereditary right M-summand of Y, then $\mathcal{A}_\ell(X)$ is weakly Morita equivalent to $\mathcal{A}_\ell(Y)$.

We conclude this section by mentioning a few properties of dual operator spaces that are invariant under weak Morita equivalence. For example, it follows from Proposition 6.5, and from facts mentioned in the last section of [**Rie**], that if X and Y are weakly Morita equivalent, and if X is type I (resp. type II, type III), then Y is type I (resp. type II, type III). If X and Y are properly infinite, then by Theorem 6.1 (iii) it follows that $X \cong Y$ completely isometrically and weak-* homeomorphically if and only if X and Y are weak column Morita equivalent. If X and Y are weakly Morita equivalent, and if all left M-projections on X commute, then Y is type I (see [**Rie**, Theorem 8.10]). Finally, it follows that if X and Y are weakly Morita equivalent, then the 'central right M-summands' of X (i.e. those corresponding to projections in the center of $\mathcal{A}_\ell(X)$) are in bijective correspondence with the 'central right M-summands' of Y. This is because weakly Morita equivalent rings have isomorphic centers. If, further, X is an operator space of the type discussed in Corollary 6.2, then so is Y.

CHAPTER 7

Central M-structure for Operator Spaces

7.1. The Operator Space Centralizer Algebra

We define the *operator space centralizer algebra* $Z(X)$ to be the subset of $CB(X)$ given by $\mathcal{A}_\ell(X) \cap \mathcal{A}_r(X)$. We remark that this space was used in the final part of [**B3**] in the study of certain operator modules, however we shall not need anything from there. Note that $\|T\| = \|T\|_{cb}$ for $T \in Z(X)$, since this is true already for $\mathcal{A}_\ell(X)$. We will see that $Z(X)$ is a slight modification of the classical Banach space centralizer algebra of X. We recall that the latter is a commutative C^*-algebra inside $B(X)$ whose projections are exactly the M-projections of X (see [**HWW**]). We will write the Banach space centralizer algebra as $\mathrm{Cent}(X)$. Although $Z(X)$ may be developed entirely analogously to the classical centralizer theory found in [**AE, Beh1, HWW**], we will instead emphasize the connections to the one-sided theory, and use these to give a swift development. The ensuing theory is in many ways less interesting than the one-sided theory presented in previous chapters, precisely because it is so close to the classical, commutative theory surveyed in [**HWW**]. We are not saying that $Z(X)$ is unimportant: in fact its projections are precisely the complete M-projections of Effros and Ruan (see [**ER1**] and 6.2 in [**BEZ**]). The corresponding complete M-ideals (or equivalently the M-ideals whose corresponding projections in the second dual are in $Z(X^{**})$) have powerful applications (see e.g. [**ER1, AR, PR**]).

From the remark at the end of Section 2.1, it follows easily that $ST = TS$ if $S \in \mathcal{A}_\ell(X), T \in \mathcal{A}_r(X)$. Also, if $T \in Z(X)$ then the involution $T^\star$ of T in $\mathcal{A}_\ell(X)$ equals its involution in $\mathcal{A}_r(X)$. To see this, write $T^\dagger$ for the latter involution. Then $T + T^\star$ and $T + T^\dagger$ are Hermitian in the Banach algebra $B(X)$. Hence their difference $T^\star - T^\dagger$ is Hermitian. Similarly, by looking at $i(T - T^\star)$ and $i(T - T^\dagger)$, we have that $i(T^\star - T^\dagger)$ is Hermitian. From the proof of Lemma A.9 we deduce that $T^\star = T^\dagger$.

Thus the subalgebra $Z(X) = \mathcal{A}_\ell(X) \cap \mathcal{A}_r(X)$ of $B(X)$ is a C^*-subalgebra of $\mathcal{A}_\ell(X)$ which is also commutative. Since the intersection of weak-* closed subspaces of a dual space is weak-* closed, it follows from Theorem 2.3 and the abstract characterization of W^*-algebras [**Sak1**] that if X is a dual operator space, then $Z(X)$ is a commutative W^*-algebra.

LEMMA 7.1. *If X is an operator space and if $P : X \to X$ is a linear idempotent map, then the following are equivalent:*

(i) $P \in Z(X)$.
(ii) P *is a complete M-projection in the sense of Effros and Ruan* [**ER1**].
(iii) $\|[P(x_{ij}) + (I-P)(y_{ij})]\| \leq \max\{\|[x_{ij}]\|, \|[y_{ij}]\|\}$ *for all $n \in \mathbb{N}$ and $[x_{ij}], [y_{ij}]$ in $M_n(X)$.*

(iv) $\max\{\|[P(x_{ij})]\|, \|[(I - P)(x_{ij})]\|\} = \|[x_{ij}]\|$, for all $n \in \mathbb{N}$ and $[x_{ij}]$ in $M_n(X)$.

PROOF. (i) $\Leftrightarrow$ (ii) Clearly the projections in $Z(X)$ are exactly the maps on X which are both a left and a right M-projection. By [**BEZ**, 6.1] these are exactly the complete M-projections.

(ii) $\Rightarrow$ (iv) This is the definition of a complete M-projection.

(iii) $\Leftrightarrow$ (iv) This is an easy exercise.

(iv) $\Rightarrow$ (i) Given (iv) we will use Theorem 2.5 (iv) to show that $P \in \mathcal{A}_\ell(X)$. Similarly $P \in \mathcal{A}_r(X)$, and then (i) follows. Note that $\|\nu_P^c(x)\|$ equals

$$\max\{\|P_{2,1}(\nu_P^c(x))\|, \|(I - P)_{2,1}(\nu_P^c(x))\|\} = \max\{\|Px\|, \|(I - P)(x)\| = \|x\|$$

for $x \in X$, and similarly for matrices. $\qquad\square$

Remark. Note that condition (i) above implies that P is a selfadjoint projection in the center of $\mathcal{A}_\ell(X)$. However the converse is false, as may be seen by looking at [**B6**, Example 6.8]. There $X = D_n P$, where D_n is the diagonal $n \times n$ matrices, and P is a suitable invertible positive matrix. Then $\mathcal{A}_\ell(X) = D_n$, which has plenty of nonzero central projections; whereas $Z(X) = \mathbb{C}$, which has only one.

COROLLARY 7.2. If X is an operator space, then $Z(X)$ is a C^*-subalgebra of the Banach space centralizer algebra $\mathrm{Cent}(X)$. Also, if X is a dual operator space, then $\mathrm{Cent}(M_\infty^w(X)) = Z(M_\infty^w(X))$, and this algebra is isomorphic to $Z(X)$ as commutative W^*-algebras.

PROOF. If X is a dual operator space, then both $Z(X)$ and $\mathrm{Cent}(X)$ are densely spanned by their projections. Since every complete M-projection is an M-projection, it follows from the last Lemma that $Z(X) \subset \mathrm{Cent}(X)$ as unital subalgebras of $B(X)$. In fact, $Z(X)$ must be a $*$-subalgebra of $\mathrm{Cent}(X)$ (as may be seen by considering Hermitians for example), and indeed it is a weak-$*$ closed subalgebra. To see this recall that the weak-$*$ topology on $\mathrm{Cent}(Z)$ is the relative weak-$*$ topology inherited from $B(X)$, and use Theorem 2.3. If $R \in \mathrm{Cent}(M_\infty^w(X))$, then we claim that R is simply the 'countably infinite amplification' $T \otimes I_\infty$ of a map $T \in \mathrm{Cent}(X)$. This is because of [**HWW**, I.3.15], which says that any operator in the centralizer commutes with every Hermitian operator. One may consider Hermitian operators on $M_\infty^w(X)$ obtained by left or right multiplying by a selfadjoint scalar matrix, and in particular Hermitian permutation and diagonal matrices (such operators are clearly Hermitian in $B(M_\infty^w(B(H)))$, if $X \subset B(H)$, and so their restrictions are Hermitian too). From such considerations we see that any $R \in \mathrm{Cent}(M_\infty^w(X))$ is of the desired form. Clearly the map $\theta : \mathrm{Cent}(M_\infty^w(X)) \to \mathrm{Cent}(X) : T \otimes I_\infty \mapsto T$ is a unital 1-1 contractive homomorphism, which forces it to be a $*$-isomorphism onto its range. It is also easy to see that θ is weak-$*$ continuous. Hence by the Krein-Smulian theorem, $\mathrm{ran}(\theta)$ is a W^*-subalgebra of $\mathrm{Cent}(X)$. On the other hand, Lemma 7.1 implies that θ takes projections in $\mathrm{Cent}(M_\infty^w(X))$ to projections in $Z(X)$. Since the span of the projections is dense, θ must map into $Z(X)$. Conversely, if $T \in Z(X)$, then by a slight modification of Proposition 5.47, we see that the amplification $T \otimes I_\infty$ is both left and right adjointable on $M_\infty^w(X)$, and hence lies in $Z(M_\infty^w(X))$. Thus $Z(M_\infty^w(X)) = \mathrm{Cent}(M_\infty^w(X))$, and θ is the desired $*$-isomorphism.

If X is a general operator space, it suffices to show that any T in the positive part of $\mathrm{ball}(Z(X))$ is also in $\mathrm{Cent}(X)$. Now T^{**} is in the positive part of ball($Z(X^{**})$) by Proposition 5.16 and its matching 'right-handed' version. Hence T^{**} is in the positive part of ball(Cent(X^{**})). The classical theory (e.g. [**HWW**, Proposition I.3.9]) implies $T \in \text{Cent}(X)$. $\qquad\square$

Note that in general $Z(X) \neq \text{Cent}(X)$. To see this one need only note that not every M-projection on an operator space is a complete M-projection (see [**ER1**]).

On the other hand, the method in the last proof yields for a general operator space X, a 1-1 $*$-homomorphism $Cent(M_\infty^w(X)) \to Cent(X)$. We shall see later that the range of this map is $Z(X)$.

LEMMA 7.3. *If X is an operator space and if $T : X \to X$ is linear, then the following are equivalent:*

   (i) T *is in the positive part of* ball($Z(X)$),
   (ii) T^{**} *is in the positive part of* ball($Z(X^{**})$),
   (iii) $\|[T(x_{ij})+(I-T)(y_{ij})]\| \leq \max\{\|[x_{ij}]\|, \|[y_{ij}]\|\}$ *for all $n \in \mathbb{N}$ and $[x_{ij}], [y_{ij}]$ in $M_n(X)$,*
   (iv) *Same as (iii) but for all matrices in $K_\infty(X)$,*
   (v) *Same as (iii) but for all matrices in $M_\infty^w(X)$.*

PROOF. First we show that (i) implies (iii)–(v). It follows by an argument towards the end of the proof of Corollary 7.2, that suitable amplifications of T are in $Z(M_n(X))$, $Z(K_\infty(X))$, and $Z(M_\infty^w(X))$, respectively. If, further, $0 \leq T \leq 1$, then these amplifications of T are easily seen to also lie between 0 and 1, and hence from Corollary 7.2 and [**HWW**, Proposition I.3.9] we obtain (iii)–(v).

(v) $\Rightarrow$ (iv) $\Rightarrow$ (iii) This is clear.

(ii) $\Rightarrow$ (i) The assumption implies that $T^{**} \in \mathcal{A}_\ell(X^{**})_{sa}$. Hence $T \in \mathcal{A}_\ell(X)_{sa}$ by Proposition 5.16. Similarly $T \in \mathcal{A}_r(X)$, and the rest is clear.

(iii) $\Rightarrow$ (ii) By taking the second dual of T, we may suppose that the hypothesis in (iii) holds for T^{**}. Hence it is easy to see that (v) holds for T^{**}. Thus by [**HWW**, Proposition I.3.9] we have $T^{**} \otimes I_\infty$ is in the positive part of the ball of Cent($M_\infty^w(X^{**})$). By the previous Lemma it follows that T^{**} is in the positive part of the ball of $Z(X^{**})$. $\qquad\square$

We now establish a host of equivalent characterizations of $Z(X)$. We briefly explain the notation in (vi) below. We recall from [**B6**] that if $(\mathcal{T}(X), J)$ is the triple envelope of X, then by one of the equivalent definitions of $\mathcal{A}_\ell(X)$ in that paper, any $T \in \mathcal{A}_\ell(X)$ has an extension $\tilde{T}$ such that $\tilde{T}J(x) = J(T(x))$ and $\tilde{T}^*J(x) = J(T^\star(x))$ for all $x \in X$. The operator $\tilde{T}$ may be viewed as an adjointable (in the C^*-module sense) operator on $\mathcal{T}(X)$; or equivalently, as an element in the multiplier algebra of the 'left C^*-algebra' $\mathcal{E}(X)$ of $\mathcal{T}(X)$.

THEOREM 7.4. *If X is an operator space and if $T : X \to X$ is linear, then the following are equivalent:*

   (i) $T \in Z(X)$.
   (ii) $T^{**} \in Z(X^{**})$.
   (iii) T *is a linear combination of four operators satisfying the condition in Lemma 7.3 (iii).*
   (iv) $T \otimes I_\infty \in \text{Cent}(M_\infty^w(X))$. *Or equivalently, T is in the range of the canonical $*$-homomorphism* Cent($M_\infty^w(X)$) $\to$ Cent(X).
   (v) *Same as (iv), with $M_\infty(X)$ replaced by $K_\infty(X)$.*

(vi) $T \in \mathcal{A}_\ell(X)$ and the associated map $\tilde{T}$ in $M(\mathcal{E}(X))$ is in the center of $M(\mathcal{E}(X))$ (see the discussion above the Theorem).

PROOF. (i) $\Rightarrow$ (ii) See the last part of the proof of Corollary 7.2.

(ii) $\Rightarrow$ (i) Since as noted above, $Z(X^{**})$ is a $*$-subalgebra of $\mathcal{A}_\ell(X^{**})$, if $T^{**} \in Z(X^{**})$ then T^{**} is a normal element of $\mathcal{A}_\ell(X^{**})$. By Proposition 5.16 (iii), $T \in \mathcal{A}_\ell(X)$. Similarly, $T \in \mathcal{A}_\mathrm{r}(X)$, so $T \in Z(X)$.

(i) $\Leftrightarrow$ (iii) This is evident from Lemma 7.3.

(iii) $\Rightarrow$ (iv) Also evident from Lemma 7.3 and [**HWW**, Proposition I.3.9].

(iv) $\Rightarrow$ (v) Follows easily for example from [**HWW**, Proposition I.3.9].

(v) $\Rightarrow$ (iv) We leave this as an exercise, using for example [**HWW**, Proposition I.3.9].

(iv) $\Rightarrow$ (i) Assuming (iv), write $T \otimes I_\infty$ as a linear combination of four positive elements R_i in the ball of $\mathrm{Cent}(M_\infty^w(X))$. By the proof of Corollary 7.2, each $R_i = S_i \otimes I_\infty$ for an $S_i \in \mathrm{Cent}(X)$. By [**HWW**, Proposition I.3.9] it is easy to see that S_i satisfies the conditions of Lemma 7.3. Thus $S_i \in Z(X)$, and hence so is T.

(vi) $\Rightarrow$ (i) Suppose that $T \in \mathcal{A}_\ell(X)$, and that $\tilde{T}$ is the associated implementing operator in the center of $M(\mathcal{E}(X))$. Now $\mathcal{T}(X)$ is a strong Morita equivalence $\mathcal{E}(X)$-$\mathcal{F}(X)$-bimodule, where $\mathcal{F}(X)$ is the 'right C^*-algebra of $\mathcal{T}(X)$'. By a basic fact about Rieffel's strong Morita equivalence (e.g. see [**BLM**, Corollary 8.1.20]), there is an operator S in the center of $M(\mathcal{F}(X))$ such that $\tilde{T}J(x) = J(T(x)) = J(x)S$ and $\tilde{T}^*J(x) = J(T^*(x)) = J(x)S^*$ for all $x \in X$. From this it is clear that $T \in \mathcal{A}_\mathrm{r}(X)$. So $T \in Z(X)$.

(i) $\Rightarrow$ (vi) Suppose that $T \in Z(X)$. Then $\tilde{T}J(x) = J(T(x))$ and $\tilde{T}^*J(x) = J(T^*(x))$ for all $x \in X$. Similarly there exists an operator S such that $J(x)S = J(T(x))$ and $J(x)S^* = J(T^*(x))$. It follows that $\tilde{T}J(x)J(y)^* = J(x)SJ(y)^* = J(x)J(y)^*\tilde{T}$. Thus it follows that $\tilde{T}$ is in the center of $M(\mathcal{E}(X))$. $\square$

Remark. The space $Z(X)$ has other reformulations in terms of structurally continuous functions, or M-boundedness, just as in the classical case [**AE, Beh1, HWW**]. This may be seen from (iv) or (v) of the last Theorem; we omit the details.

COROLLARY 7.5. Let X be an operator space. Then $Z(X) \cong \mathrm{Cent}(M_\infty^w(X)) \cong \mathrm{Cent}(K_\infty(X))$ as commutative C^*-algebras.

PROPOSITION 7.6.   (i) If X is a Banach space, $\mathrm{Cent}(X) = Z(\mathrm{MIN}(X))$.

(ii) If $\mathcal{A}$ is a C^*-algebra then $\mathrm{Cent}(\mathcal{A}) = Z(\mathcal{A})$, and this also equals the center of the multiplier C^*-algebra $M(\mathcal{A})$.

(iii) If $\mathcal{A}$ is an operator algebra with a contractive approximate identity, then $\mathrm{Cent}(\mathcal{A}) = Z(\mathcal{A})$, and this also equals the diagonal of the center of $M(\mathcal{A})$.

PROOF. (i) This follows from [**B6**, Corollary 4.22].

(ii) This is well known, and in any case follows from (iii).

(iii) First suppose that $\mathcal{A}$ is a unital dual operator algebra (a unital weak-$*$ closed subalgebra of a W^*-algebra). Then $M(\mathcal{A}) = \mathcal{A}$, and the three algebras which we are trying to show are equal, are at least commutative W^*-algebras, and may all be viewed as subalgebras of $B(\mathcal{A})$. They are therefore spanned by their projections. If P is an M-projection on $\mathcal{A}$, then from [**ER2**] we have $Px = ex$ for a central projection $e \in \mathcal{A}$, so that P is a complete M-projection. Thus the first two algebras are equal. View $Z(\mathcal{A}) \subset \mathcal{A}_\ell(\mathcal{A}) \subset CB(\mathcal{A})$. We employ the canonical

*-homomorphism from $\mathcal{A}_\ell(\mathcal{A})$ into $\mathcal{A}$ given by $T \mapsto T(1)$, which maps onto the 'diagonal' $\mathcal{A} \cap \mathcal{A}^*$. The restriction θ of this *-homomorphism to $Z(\mathcal{A})$ takes the M-projection $Px = ex$ above to the central projection $e \in \mathcal{A}$. By linearity and density considerations, θ maps into the diagonal of the center of $\mathcal{A}$, and has dense range. Thus θ is a *-isomorphism of $Z(\mathcal{A})$ onto the diagonal of the center of $\mathcal{A}$.

Next suppose that $\mathcal{A}$ is not a dual algebra. If $T \in \mathrm{Cent}(\mathcal{A})$, then $T^{**} \in \mathrm{Cent}(\mathcal{A}^{**})$, and hence by the last paragraph there exists a unique η in the diagonal of the center of $\mathcal{A}^{**}$ with $T^{**}(\nu) = \eta\nu = \nu\eta$ for all $\nu \in \mathcal{A}^{**}$. Since $T^{**}(a) = T(a) \in \mathcal{A}$ for $a \in \mathcal{A}$, it follows that η is in $M(\mathcal{A})$, and indeed is in the center of that algebra. The mapping $\pi : \mathrm{Cent}(\mathcal{A}) \to Z(M(\mathcal{A})) : T \mapsto \eta$ is a 1-1 isometric homomorphism. Thus π is a *-monomorphism and maps into the diagonal of the center of $M(\mathcal{A})$. Conversely, if η is a selfadjoint (Hermitian) element in the center of $M(\mathcal{A})$, then we may regard η as a selfadjoint (Hermitian) in $\mathcal{A}^{**}$ with $\eta a = a\eta$ for all $a \in \mathcal{A}$. It follows that η is in the diagonal of the center of $\mathcal{A}^{**}$, so that by the last paragraph there is a selfadjoint $S \in Z(\mathcal{A}^{**})$ with $S(\nu) = \eta\nu$ for all $\nu \in \mathcal{A}^{**}$. Since $S(a) = \eta a \in \mathcal{A}$ for all $a \in \mathcal{A}$, it follows that $S \in Z(\mathcal{A})$. Hence π is surjective, and is a *-isomorphism, and the three algebras coincide as desired.   □

Remark. Another useful situation when $Z(X) = Cent(X)$ is if X is a Hilbert C^*-module or TRO (ternary ring of operators). Moreover, as is essentially well known, the M-ideals in such a space X coincide with the complete M-ideals. Indeed, if X is a TRO then these M-ideals are exactly the closed XX^*-X^*X-submodules; whereas if X is a Hilbert 'C^*-bimodule' over two C^*-algebras A and B, then these M-ideals are exactly the closed A-B-subbimodules. See e.g. 8.5.20 in [**BLM**] for proofs.

It then follows from the the fourth 'bullet' in the list towards the end of Section 2.3, that the M-summands of such X, are exactly the complete M-summands. Thus every M-projection on X^{**} is in $Z(X^{**})$, and so $Cent(X^{**}) = Z(X^{**})$. If $T \in Cent(X)$, then $T^{**} \in Cent(X^{**}) = Z(X^{**})$, so that $T \in Z(X)$ by Theorem 7.4. Thus $Cent(X) = Z(X)$.

We will not attempt to systematically transfer results from earlier chapters to the case of $Z(X)$ and complete M-ideals. This is indeed quite routine, and almost identical to the classical case. We do observe that complete M-projections do not exhibit the bad behavior that we saw in Example 5.4. That is, if P is a projection in $Z(X)$, and if $Y = P(X)$, then $P\mathcal{M}_\ell(X)P = \mathcal{M}_\ell(Y), P\mathcal{A}_\ell(X)P = \mathcal{A}_\ell(Y)$, and $PZ(X) = Z(Y)$. This is because in this case $X \cong P(X) \oplus_\infty (I - P)(X)$, and one may appeal for example to [**B6**, Theorem A.13] and the remark after it.

7.2. Complete L-Projections and the Cunningham Algebra

It is clear that P is a complete L-projection in the sense of [**ER1**] if and only if P^* is a complete M-projection. One may define the operator space version of the classical Cunningham algebra to be the closed linear span in $B(X)$ of the complete L-projections on X. The theory of the operator space Cunningham algebra is almost identical to that of the classical Cunningham algebra (see e.g. [**Beh1, HWW**]). Because of this, and since we have no applications for it, we leave the details to the interested reader. Indeed the operator space Cunningham algebra of an operator space X is a C^*-subalgebra of the classical Cunningham algebra of X. Also, it is *-isomorphic to $Z(X^*)$, and it may also be identified with the classical Cunningham algebra of $S^1_\infty(X)$, where the latter is defined to be the operator space projective tensor product of the predual of M_∞ with X.

CHAPTER 8

Future directions

We begin by mentioning two of the most important remaining unsolved problems in the basic theory of one-sided M-ideals. Most importantly, we do not have a characterizations of one-sided M-ideals and summands in terms of ball intersection properties, in the spirit of [**AE**] and later researchers (see the works of R. Evans, O. Hustad, A. Lima, and others cited in [**HWW**]). Probably related to this is the question of proximinality of one-sided M-ideals. We recall that $J \subset X$ is proximinal if for all $x \in X$, the distance $d(x, J)$ is achieved. We suspect that there are examples of one-sided M-ideals which are not proximinal, in which case one could ask for sufficient conditions for proximinality. Such criteria may well be very useful in operator space theory; for example one may hope to find a connection to Kirchberg's deep proximinality results from e.g. [**Kir**]. In fact, there probably are nonproximinal left ideals with contractive right approximate identities in a unital operator algebra. Such ideals are left M-ideals, as we stated in Section 4.5. In particular, we imagine that some of the examples discussed at the end of that section are nonproximinal. For some related discussion of proximinality in C^*- and operator algebras see [**DP**].

In this memoir, we have only attempted to transfer the 'basic theory' of M-ideals to the one-sided situation. The more advanced theory treated in later chapters of [**HWW**] remains wide open—for example, the theory of spaces that are one-sided M-ideals in their bidual. We will however wait until applications arise that call for the development of such further theory.

The first author has recently introduced one-sided multipliers between two different spaces [**B4**]. We would guess that the proofs of several of our results for multipliers in this article (for example the polar decomposition theory) should adapt to give analogous results and applications for the new class of multipliers. In a similar spirit, there is a recent notion of 'quasimultipliers' introduced by Kaneda and Paulsen [**Kan**]. Quasimultipliers are intimately related to the one-sided multipliers, and may be developed in an analogous fashion. Kaneda has suggested a 'quasi'- or 'inner' version of one-sided M-ideals which would be interesting to pursue. We would guess that the proofs of several of our results for multipliers and ideals should adapt to give analogous results for quasimultipliers and 'quasi-M-ideals'. However since the quasimultiplier space has so little structure in general, others of our results may fail, or be quite difficult, to transfer.

APPENDIX A

Some Results from Banach Space Theory

In this appendix, we collect some facts that we shall need from Banach space theory. We provide references when possible, and proofs otherwise.

A.1. Dual spaces

THEOREM A.1. (Krein-Smulian)

(i) *Let X be a dual Banach space, and Y be a linear subspace of X. Then Y is weak-* closed in X if and only if $\mathrm{ball}(Y)$ is closed in the weak-* topology on X. In this case, Y is also a dual Banach space, with predual $X_*/^\perp Y$, and the inclusion of Y into X is weak-* continuous.*

(ii) *A bounded linear map T between dual Banach spaces is weak-* continuous if and only if whenever $x_i \to x$ is a bounded net converging weak-* in the domain space, it follows that $T(x_i) \to T(x)$ weak-*.*

(iii) *Let X and Y be dual Banach spaces, and $T : X \to Y$ be a weak-* continuous linear isometry. Then T has weak-* closed range V say, and T is a weak-* homeomorphism of X onto V.*

Variations of the Krein-Smulian theorem can be found in any functional analysis book. The above formulation and its proof appears in [**ER5, B3**].

LEMMA A.2. *Let X be a Banach space and J be a closed linear subspace of X. If $J^{\perp\perp}$ is contractively complemented in X^{**} and J is a dual Banach space, then J is contractively complemented in X.*

PROOF. Let Y be a Banach space such that there exists an isometric isomorphism $\Phi : Y^* \to J$. Let $P : X^{**} \to J^{\perp\perp}$ be a contractive surjection, and let $\Psi : J^{\perp\perp} \to J^{**}$ be the canonical isometric isomorphism. For any Banach space Z, let $\iota_Z : Z \to Z^{**}$ be the canonical isometric inclusion. Then the composition

$$X \xrightarrow{\iota_X} X^{**} \xrightarrow{P} J^{\perp\perp} \xrightarrow{\Psi} J^{**} \xrightarrow{(\Phi^{-1})^{**}} Y^{***} \xrightarrow{\iota_Y^*} Y^* \xrightarrow{\Phi} J$$

is a contractive surjection. $\qquad\square$

LEMMA A.3. *Let X be a Banach space and let J_a, $a \in A$, be closed linear subspaces of X. Then*

(i) $\left(\bigcup_{a \in A} J_a \right)^\perp = \bigcap_{a \in A} J_a^\perp$.

(ii) $\left(\bigcap_{a \in A} J_a \right)^\perp = \overline{\mathrm{span}}^{\mathrm{wk}^*} \left\{ \bigcup_{a \in A} J_a^\perp \right\}$.

PROOF. (i) This is clear. In fact, the J_a's can be arbitrary sets in this case.

(ii) Likewise, the inclusion $\overline{\mathrm{span}}^{\mathrm{wk}^*} \left\{ \bigcup_{a \in A} J_a^\perp \right\} \subset \left(\bigcap_{a \in A} J_a \right)^\perp$ is straightforward, even if the J_a's are arbitrary sets. Now suppose that $f \in \left(\bigcap_{a \in A} J_a \right)^\perp$ but $f \notin \overline{\mathrm{span}}^{\mathrm{wk}^*} \left\{ \bigcup_{a \in A} J_a^\perp \right\}$. Then there exists a weak-* continuous linear functional

Φ on X^* such that $\Phi(f) \neq 0$ but $\Phi(g) = 0$ for all $g \in \overline{\operatorname{span}}^{\operatorname{wk}^*} \{\bigcup_{a \in A} J_a^\perp\}$. Since $(X^*, \operatorname{wk}^*)^* = X$, there exists an $x \in X$ such that $f(x) \neq 0$ but $g(x) = 0$ for all $g \in \overline{\operatorname{span}}^{\operatorname{wk}^*} \{\bigcup_{a \in A} J_a^\perp\}$. It follows that $x \notin \bigcap_{a \in A} J_a$, so that there exists a $b \in A$ with $x \notin J_b$. Because $^\perp(J_b^\perp) = \overline{\operatorname{span}}\{J_b\} = J_b$, there exists a $g \in J_b^\perp$ such that $g(x) \neq 0$, a contradiction. $\square$

LEMMA A.4. *Let X be a Banach space and let J_a, $a \in A$, be weak-* closed linear subspaces of X^*. Then*

   (i) $^\perp(\bigcup_{a \in A} J_a) = \bigcap_{a \in A} {}^\perp J_a$.
   (ii) $^\perp(\bigcap_{a \in A} J_a) = \overline{\operatorname{span}}\{\bigcup_{a \in A} {}^\perp J_a\}$

PROOF. (i) This is clear for arbitrary sets J_a.

(ii) The inclusion $\overline{\operatorname{span}}\{\bigcup_{a \in A} {}^\perp J_a\} \subset {}^\perp(\bigcap_{a \in A} J_a)$ is easy, even for arbitrary sets J_a. Now suppose that $x \in {}^\perp(\bigcap_{a \in A} J_a)$ but $x \notin \overline{\operatorname{span}}\{\bigcup_{a \in A} {}^\perp J_a\}$. Then there exists an $f \in X^*$ such that $f(x) \neq 0$ but $f(y) = 0$ for all $y \in \overline{\operatorname{span}}\{\bigcup_{a \in A} {}^\perp J_a\}$. It follows that $f \notin \bigcap_{a \in A} J_a$, so that $f \notin J_b$ for some $b \in A$. Since $(^\perp J_b)^\perp = \overline{\operatorname{span}}^{\operatorname{wk}^*}\{J_b\} = J_b$, there exists a $y \in {}^\perp J_b$ such that $f(y) \neq 0$, a contradiction. $\square$

LEMMA A.5 ([**HWW**], Lemma I.1.14). *Let X be a Banach space, and J and K be closed linear subspaces of X. Then the following are equivalent:*

   (i) $J + K$ *is norm closed in X.*
   (ii) $J^\perp + K^\perp$ *is norm closed in X^*.*
   (iii) $J^\perp + K^\perp$ *is weak-* closed in X^*.*

LEMMA A.6 ([**Fab**], Exercise 5.13). *Let X be a Banach space, and J and K be closed linear subspaces of X. Then the following are equivalent:*

   (i) $X = J \oplus K$.
   (ii) $X^* = J^\perp \oplus K^\perp$.
   (iii) $X^{**} = J^{\perp\perp} \oplus K^{\perp\perp}$.

LEMMA A.7 ([**Fab**], Exercise 5.18). *Let X be a Banach space and $P : X^* \to X^*$ be a projection. If the range and kernel of P are weak-* closed, then P is weak-* continuous.*

A.2. Miscellaneous facts

LEMMA A.8 ([**Fab**], Exercise 5.27). *Let X be a Banach space, and Y and Z be closed linear subspaces of X. If Y is finite-dimensional, then $Y + Z$ is closed.*

LEMMA A.9. *Let $\mathcal{B}$ be a unital Banach algebra, and $a, b, c, d \in \operatorname{Her}(\mathcal{B})$. If $a + ib = c + id$, then $a = c$ and $b = d$.*

PROOF. It suffices to show that if $a, b \in \operatorname{Her}(\mathcal{B})$ and $a = ib$, then $a = b = 0$. For such a, b, and for any state f on $\mathcal{B}$, we have $f(a) = f(b) = 0$. Thus the *numerical radii* are zero, and it is well known that this implies that $a = b = 0$ (see e.g. [**BD**]). $\square$

LEMMA A.10. *Let X be a Banach space and $P, Q : X \to X$ be bounded projections. If $\|PQ\| < 1$, then $\operatorname{ran}(P) + \operatorname{ran}(Q)$ is closed.*

PROOF. We give Akemann's argument. Assume that $\operatorname{ran}(P) + \operatorname{ran}(Q)$ is not closed. Consider the bounded linear map

$$T : \operatorname{ran}(P) \oplus_\infty \operatorname{ran}(Q) \to X : (x, y) \mapsto x + y.$$

Since $\text{ran}(T) = \text{ran}(P) + \text{ran}(Q)$, T cannot be bounded below. Therefore, there exist sequences (x_n) in $\text{ran}(P)$ and (y_n) in $\text{ran}(Q)$ such that $\|x_n\| = \|y_n\| = 1$ and $\|x_n - y_n\| \leq 2^{-n}$ for all $n \in \mathbb{N}$. But then

$$1 = \|x_n\| = \|Px_n\| \leq \|P(x_n - y_n)\| + \|Py_n\| \leq \|P\|\|x_n - y_n\| + \|PQy_n\|$$

and this is dominated by $2^{-n}\|P\| + \|PQ\|$, for all $n \in \mathbb{N}$. Choosing n sufficiently large results in a contradiction. $\square$

Infinite Matrices over an Operator Space

In this appendix we briefly discuss the theory of infinite matrices over an operator space. For a more detailed discussion, see for example [**ER4**, Section 10.1] or [**BLM**, 1.2.26,1.6.3,2.2.3], and the references contained therein.

Let X be an operator space and I be a (typically infinite) index set. By $M_I^w(X)$ we mean the linear space of all $I \times I$ matrices

$$x = \begin{bmatrix} & \vdots & \\ \cdots & x_{ij} & \cdots \\ & \vdots & \end{bmatrix}_{i,j \in I}$$

over X such that

$$\|x\| \equiv \sup \left\{ \left\| \begin{bmatrix} & \vdots & \\ \cdots & x_{ij} & \cdots \\ & \vdots & \end{bmatrix}_{i,j \in F} \right\| : F \subset I \text{ is finite} \right\} < \infty.$$

It is not hard to verify that $M_I^w(X)$ is a Banach space. In fact, it is an operator space with respect to the matrix norms determined by the linear isomorphisms

$$M_n(M_I^w(X)) = M_I^w(M_n(X)).$$

Similar definitions and observations pertain to 'rectangular' matrices $M_{I,J}^w(X)$. We write $C_I^w(X)$ for $M_{I,1}^w(X)$, and $R_I^w(X)$ for $M_{1,I}^w(X)$. In fact $M_{I,J}^w(X) \cong R_J^w(C_I^w(X)) \cong C_I^w(R_J^w(X))$. If $I = \aleph_0$ we simply write $C_\infty^w(X)$ and $R_\infty^w(X)$ for $C_I^w(X)$ and $R_I^w(X)$. If $X = \mathbb{C}$, then we write $C_I, R_I, C_\infty, R_\infty$ for the above spaces; these are known as *column and row Hilbert spaces*.

Some simple examples and observations are in order:

- $M_I^w(\mathbb{C}) = B(\ell^2(I))$. This space is typically abbreviated M_I.
- More generally, if $\mathcal{R}$ is a W^*-algebra, then $M_I^w(\mathcal{R}) = M_I \overline{\otimes} \mathcal{R}$. In particular, $M_I^w(\mathcal{R})$ is again a W^*-algebra.
- If $\mathcal{A}$ is C^*-algebra, then $M_I^w(\mathcal{A})$ need not be a C^*-algebra. In fact, it need not even be an algebra.

Contained inside $M_I^w(X)$ are a number of distinguished operator subspaces:

- $M_I(X) = M_I \check{\otimes} X$, the norm closed linear span of $\{A \otimes x : A \in M_I, x \in X\}$ (the 'elementary tensors').
- $K_I(X)$, the norm closure of the 'finite rank' elements of $M_I^w(X)$.
- $C_I^w(X)$ and $R_I^w(X)$, which may be identified with a fixed column and row of $M_I^w(X)$, respectively.

- $C_I(X)$ and $R_I(X)$, which may be identified with a fixed column and row of $M_I(X)$, respectively. Equivalently, with a fixed column and row of $K_I(X)$, respectively.
- $R_I(C_I^w(X))$, $C_I^w(R_I(X))$, $C_I(R_I^w(X))$, and $R_I^w(C_I(X))$.

The reader will have no trouble verifying the following diagram of inclusions:

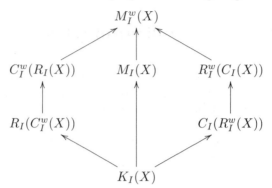

And of course $C_I(X) \subset C_I^w(X)$ and $R_I(X) \subset R_I^w(X)$. We write ∞ in place of the subscript I if I is countably infinite; thus for example $K_\infty(X) \cong C_\infty(R_\infty(X))$.

Some more examples and observations:

- Taking $X = \mathbb{C}$ in the 'column and row matrix' definitions above, we have $C_I^w = C_I$ and $R_I^w = R_I$. Indeed if H is a Hilbert space with orthonormal basis $\{e_i : i \in I\}$, then $H_c = C_I$ completely isometrically and $H_r = R_I$ completely isometrically.
- If $\mathcal{A}$ is a unital operator algebra, then $C_I(R_I^w(\mathcal{A}))$ and $R_I(C_I^w(\mathcal{A}))$ are operator algebras with contractive approximate identities, and $R_I^w(C_I(\mathcal{A}))$ and $C_I^w(R_I(\mathcal{A}))$ are unital operator algebras.
- The second dual of $K_I(X)$ is $M_I^w(X^{**})$ (see [**ER4**, Theorem 10.1.4]). The second dual of $C_I(X)$ is $C_I^w(X^{**})$, and a similar assertion holds for rows.

As a final remark, we note that if X is an operator space and I is an index set, then $M_I^w(X^*)$ is a dual operator space. Indeed, $M_I^w(X^*) = CB(X, M_I)$ completely isometrically. A bounded net (f_α) in $CB(X, M_I)$ converges to $f \in CB(X, M_I)$ in the weak-* topology if and only if $f_\alpha(x) \to f(x)$ weak-* for all $x \in X$, which is the same as saying that $\operatorname{Tr}(f_\alpha(x)A) \to \operatorname{Tr}(f(x)A)$ for all $x \in X$ and all trace-class operators $A \in M_I$. This, in turn, is the same as saying that $f_\alpha(x)_{i,j} \to f(x)_{i,j}$ for all $x \in X$ and all $i, j \in I$. Therefore, weak-* convergence of bounded nets in $M_I^w(X^*)$ is the same as entry-wise weak-* convergence.

Bibliography

[Ake1] C. Akemann, *The general Stone-Weierstrass problem*, J. Funct. Anal. **4** (1969), 277–294.

[Ake2] _____, *Left ideal structure of C^*-algebras*, J. Funct. Anal. **6** (1970), 305–317.

[AE] E. M. Alfsen and E. G. Effros, *Structure in real Banach spaces. I, II*, Ann. of Math. **96** (1972), 98–173.

[AR] A. Arias and H. P. Rosenthal, *M-complete approximate identities in operator spaces*, Studia Math. **141** (2000), 143–200.

[Arv1] W. B. Arveson, *Subalgebras of C^*-algebras*, Acta Math. **123** (1969), 141–224.

[Arv2] _____, *Subalgebras of C^*-algebras. II*, Acta Math. **128** (1972), 271–308.

[Beh1] E. Behrends, *M-structure and the Banach-Stone theorem*, Lecture Notes in Mathematics 736, Springer-Verlag, Berlin-Heidelberg-New York, 1979.

[Beh2] _____, *Normal operators and multipliers on complex Banach spaces and a symmetry property of L^1-predual spaces*, Israel J. Math. **47** (1984), 23–28.

[B1] D. P. Blecher, *A completely bounded characterization of operator algebras*, Math. Ann. **303** (1995), 227–239.

[B2] _____, *Geometry of the tensor product of C^*-algebras*, Math. Proc. Camb. Phil. Soc. **104** (1988), 119–127.

[B3] _____, *Multipliers and dual operator algebras*, J. Funct. Anal. **183** (2001), 498–525.

[B4] _____, *Multipliers, C^*-modules, and algebraic structure in spaces of Hilbert space operators*, in "Operator Algebras, Quantization, and Non-Commutative Geometry", 85–128, Contemp. Math., 365, Amer. Math. Soc., Providence, RI, 2004.

[B5] _____, *One-sided ideals and approximate identities in operator algebras*, J. Austral. Math. Soc. **76** (2004), 425–447.

[B6] _____, *The Shilov boundary of an operator space and the characterization theorems*, J. Funct. Anal. **182** (2001), 280–343.

[BEZ] D. P. Blecher, E. G. Effros, and V. Zarikian, *One-sided M-ideals and multipliers in operator spaces. I*, Pacific J. Math. **206** (2002), 287–319.

[BLM] D. P. Blecher and C. Le Merdy, *Operator algebras and their modules–an operator space approach*, Oxford University Press, 2004.

[BM] D. P. Blecher and B. Magajna, *Duality and operator algebras I & II*, To appear, J. Funct. Anal.

[BMP] D. P. Blecher, P. S. Muhly, and V. I. Paulsen, *Categories of operator modules (Morita equivalence and projective modules)* Mem. Amer. Math. Soc. **143** (2000), no. 681.

[BP] D. P. Blecher and V. I. Paulsen, *Multipliers of operator spaces and the injective envelope*, Pacific J. Math. **200** (2001), 1–17.

[BSZ] D. P. Blecher, R. R. Smith, and V. Zarikian, *One-sided projections on C^*-algebras*, J. Operator Theory **51** (2004), 201–220.

[BZ] D. P. Blecher and V. Zarikian, *Multiplier operator algebras and applications*, Proceedings of the National Academy of Sciences of the U.S.A. **101** (2004), 727–731.

[Boh] F. Bohnenblust, *A characterization of complex Hilbert spaces*, Portugaliae Math. **3** (1942), 103–109.

[BD] F. F. Bonsall and J. Duncan, *Complete normed algebras*, Springer-Verlag, New York-Heidelberg, 1973.

[DP] K. R. Davidson and S. C. Power, *Best approximation in C^*-algebras*, J. Reine Angew. Math. **368** (1986), 43–62.

[Eff1] E. G. Effros, *Order ideals in a C^*-algebra and its dual*, Duke Math. J. **30** (1963), 391–412.

[EK] E. G. Effros and A. Kishimoto, *Module maps and Hochschild-Johnson cohomology*, Indiana Univ. Math. J. **36** (1987), 257–276.

[ER1]   E. G. Effros and Z-J. Ruan, *Mapping spaces and liftings for operator spaces*, Proc. London Math. Soc. **69** (1994), 171–197.

[ER2]   _____, *On non-self-adjoint operator algebras*, Proc. Amer. Math. Soc. **110** (1990), 915–922.

[ER3]   _____, *Operator space tensor products and Hopf convolution algebras*, J. Operator Theory, **50** (2003), 131-156.

[ER4]   _____, *Operator spaces*, Oxford University Press, Oxford, 2000.

[ER5]   _____, *Representations of operator bimodules and their applications*, J. Operator Theory **19** (1988), 137–157.

[Fab]   Fabian et al., *Functional analysis and infinite-dimensional geometry*, Springer-Verlag, New York, 2001.

[GKS]   G. Godefroy, N. Kalton, and P. Saphar, *Unconditional ideals in Banach spaces*, Studia Math. **104** (1993), 13–59.

[Hal]   P. Halmos, *Introduction to Hilbert space*, Chelsea Publishing Company, New York, 1957.

[Ham]   M. Hamana, *Triple envelopes and Silov boundaries of operator spaces*, Math. J. Toyama University **22** (1999), 77–93.

[HWW]   P. Harmand, D. Werner, and W. Werner, *M-ideals in Banach spaces and Banach algebras*, Lecture Notes in Mathematics 1547, Springer-Verlag, Berlin-Heidelberg-New York, 1993.

[JP]   M. Junge and J. Parcet, *The norm of sums of independent noncommutative random variables in $L_p(\ell_1)$*, Preprint (2004).

[KR1]   R. V. Kadison and J. R. Ringrose, *Fundamentals of the theory of operator algebras. Vol. I: Elementary theory*, Amer. Math. Soc., Providence, 1997.

[KR2]   R. V. Kadison and J. R. Ringrose, *Fundamentals of the theory of operator algebras. Vol. II: Advanced theory*, Amer. Math. Soc., Providence, 1997.

[Kak]   S. Kakutani, *Some characterizations of Euclidean space*, Japan. J. Math. **16** (1939), 93–97.

[Kan]   M. Kaneda, *Multipliers and algebrizations of operator spaces*, Ph.D. thesis, University of Houston, 2003.

[KP]   M. Kaneda and V. I. Paulsen, *Quasimultipliers of operator spaces*, J. Funct. Anal. 217 (2004), no. 2, 347–365.

[Kir]   E. Kirchberg, *On restricted perturbations in inverse images and a description of normalizer algebras in C^*-algebras*, J. Funct. Anal. **129** (1995), 35–63.

[Lan]   E. C. Lance, *Hilbert C^*-modules–A toolkit for operator algebraists*, London Math. Soc. Lecture Note Series, Cambridge University Press, Cambridge, 1995.

[Lim]   A. Lima, *The metric approximation property, norm-one projections and intersection properties of balls*, Israel J. Math. **84** (1993), 451–475.

[RN]   M. Neal and B. Russo, *State spaces of JB^*-triples*, Math. Ann. **328** (2004), 585-624.

[Pas]   W. L. Paschke, *Inner product modules over B^*-algebras*, Trans. Amer. Math. Soc. **182** (1973), 443-468.

[Pau]   V. Paulsen, *Completely bounded maps and operator algebras*, Cambridge University Press, Cambridge, 2002.

[Pis1] G. Pisier, *Introduction to operator space theory*, London Math. Soc. Lecture Note Series 294, Cambridge University Press, Cambridge, 2003.

[Pis2] _____, *The operator Hilbert space OH, complex interpolation and tensor norms*, Mem. Amer. Math. Soc. **122** (1996), no. 585, 1–103.

[PR]   Y-T. Poon and Z-J. Ruan, *Operator algebras with contractive approximate identities*, Canadian J. Math. **46** (1994), 397–414.

[Pow]   R. T. Powers, *Representations of uniformly hyperfinite algebras and their associated von Neumann rings*, Ann. of Math. **86** (1967), 138–171.

[Pro]   R. T. Prosser, *On the ideal structure of operator algebras*, Mem. Amer. Math. Soc. **45** (1963).

[Rie]   M. A. Rieffel, *Morita equivalence for C^*-algebras and W^*-algebras*, J. Pure Appl. Algebra **5** (1974), 51–96.

[Rua]   Z-J. Ruan, *Type decomposition and rectangular AFD property for W^*-TRO's*, Canadian J. Math. **56** (2004), 843-870.

[Sak1] S. Sakai, *A characterization of W^*-algebras*, Pacific J. Math. **6** (1956), 763–773.

[Sak2] _____, *C^*-algebras and W^*-algebras*, Springer-Verlag, Berlin, 1971.

[SW] R. Smith and J. Ward, *M-ideal structure in Banach algebras*, J. Funct. Anal. **27** (1978), 337–349.

[vN] J. von Neumann, *Functional Operators, Volume II: The Geometry of Orthogonal Spaces*, Annals of Mathematics Studies 22, Princeton University Press, Princeton, 1950.

[Weg] N. E. Wegge-Olsen, *K-theory and C*-algebras*, Oxford University Press, Oxford, 1993.

[KWer] K. H. Werner, *A characterization of C*-algebras by nh-projections on matrix ordered spaces*, unpublished preprint, Saarbrucken.

[WWer] W. Werner, *Multipliers on matrix ordered operator spaces and some K-groups*, J. Funct. Anal. **206** (2004), 356–378.

[Wil] G. Willis, *Factorization in finite codimensional ideals of group algebras,* Proc. London Math. Soc. (3) 82 (2001), 676–700.

[Z1] V. Zarikian, *Complete one-sided M-ideals in operator spaces*, Ph.D. thesis, UCLA, 2001.

[Z2] _____ , *Local characterizations of one-sided M-ideals (working title)*, in preparation.

Editorial Information

To be published in the *Memoirs*, a paper must be correct, new, nontrivial, and significant. Further, it must be well written and of interest to a substantial number of mathematicians. Piecemeal results, such as an inconclusive step toward an unproved major theorem or a minor variation on a known result, are in general not acceptable for publication. Papers appearing in *Memoirs* are generally at least 80 and not more than 200 published pages in length. Papers less than 80 or more than 200 published pages require the approval of the Managing Editor of the Transactions/Memoirs Editorial Board.

As of September 30, 2005, the backlog for this journal was approximately 14 volumes. This estimate is the result of dividing the number of manuscripts for this journal in the Providence office that have not yet gone to the printer on the above date by the average number of monographs per volume over the previous twelve months, reduced by the number of volumes published in four months (the time necessary for preparing a volume for the printer). (There are 6 volumes per year, each containing at least 4 numbers.)

A Consent to Publish and Copyright Agreement is required before a paper will be published in the *Memoirs*. After a paper is accepted for publication, the Providence office will send a Consent to Publish and Copyright Agreement to all authors of the paper. By submitting a paper to the *Memoirs*, authors certify that the results have not been submitted to nor are they under consideration for publication by another journal, conference proceedings, or similar publication.

Information for Authors

Memoirs are printed from camera copy fully prepared by the author. This means that the finished book will look exactly like the copy submitted.

The paper must contain a *descriptive title* and an *abstract* that summarizes the article in language suitable for workers in the general field (algebra, analysis, etc.). The *descriptive title* should be short, but informative; useless or vague phrases such as "some remarks about" or "concerning" should be avoided. The *abstract* should be at least one complete sentence, and at most 300 words. Included with the footnotes to the paper should be the 2000 *Mathematics Subject Classification* representing the primary and secondary subjects of the article. The classifications are accessible from `www.ams.org/msc/`. The list of classifications is also available in print starting with the 1999 annual index of *Mathematical Reviews*. The Mathematics Subject Classification footnote may be followed by a list of *key words and phrases* describing the subject matter of the article and taken from it. Journal abbreviations used in bibliographies are listed in the latest *Mathematical Reviews* annual index. The series abbreviations are also accessible from `www.ams.org/publications/`. To help in preparing and verifying references, the AMS offers MR Lookup, a Reference Tool for Linking, at `www.ams.org/mrlookup/`. When the manuscript is submitted, authors should supply the editor with electronic addresses if available. These will be printed after the postal address at the end of the article.

Electronically prepared manuscripts. The AMS encourages electronically prepared manuscripts, with a strong preference for $\mathcal{A}\mathcal{M}\mathcal{S}$-LATEX. To this end, the Society has prepared $\mathcal{A}\mathcal{M}\mathcal{S}$-LATEX author packages for each AMS publication. Author packages include instructions for preparing electronic manuscripts, the *AMS Author Handbook*, samples, and a style file that generates the particular design specifications of that publication series. Though $\mathcal{A}\mathcal{M}\mathcal{S}$-LATEX is the highly preferred format of TEX, author packages are also available in $\mathcal{A}\mathcal{M}\mathcal{S}$-TEX.

Authors may retrieve an author package from e-MATH starting from `www.ams.org/tex/` or via FTP to `ftp.ams.org` (login as `anonymous`, enter username as password, and type `cd pub/author-info`). The *AMS Author Handbook* and the *Instruction Manual* are available in PDF format following the author packages link from `www.ams.org/tex/`. The author package can be obtained free of charge by sending email to pub@ams.org (Internet) or from the Publication Division, American Mathematical Society, 201 Charles St., Providence, RI 02904, USA. When requesting an author package, please specify $\mathcal{AMS}$-LaTeX or $\mathcal{AMS}$-TeX, Macintosh or IBM (3.5) format, and the publication in which your paper will appear. Please be sure to include your complete mailing address.

Sending electronic files. After acceptance, the source file(s) should be sent to the Providence office (this includes any TeX source file, any graphics files, and the DVI or PostScript file).

Before sending the source file, be sure you have proofread your paper carefully. The files you send must be the EXACT files used to generate the proof copy that was accepted for publication. For all publications, authors are required to send a printed copy of their paper, which exactly matches the copy approved for publication, along with any graphics that will appear in the paper.

TeX files may be submitted by email, FTP, or on diskette. The DVI file(s) and PostScript files should be submitted only by FTP or on diskette unless they are encoded properly to submit through email. (DVI files are binary and PostScript files tend to be very large.)

Electronically prepared manuscripts can be sent via email to pub-submit@ams.org (Internet). The subject line of the message should include the publication code to identify it as a Memoir. TeX source files, DVI files, and PostScript files can be transferred over the Internet by FTP to the Internet node e-math.ams.org (130.44.1.100).

Electronic graphics. Comprehensive instructions on preparing graphics are available at www.ams.org/jourhtml/graphics.html. A few of the major requirements are given here.

Submit files for graphics as EPS (Encapsulated PostScript) files. This includes graphics originated via a graphics application as well as scanned photographs or other computer-generated images. If this is not possible, TIFF files are acceptable as long as they can be opened in Adobe Photoshop or Illustrator. No matter what method was used to produce the graphic, it is necessary to provide a paper copy to the AMS.

Authors using graphics packages for the creation of electronic art should also avoid the use of any lines thinner than 0.5 points in width. Many graphics packages allow the user to specify a "hairline" for a very thin line. Hairlines often look acceptable when proofed on a typical laser printer. However, when produced on a high-resolution laser imagesetter, hairlines become nearly invisible and will be lost entirely in the final printing process.

Screens should be set to values between 15% and 85%. Screens which fall outside of this range are too light or too dark to print correctly. Variations of screens within a graphic should be no less than 10%.

Inquiries. Any inquiries concerning a paper that has been accepted for publication should be sent directly to the Electronic Prepress Department, American Mathematical Society, 201 Charles St., Providence, RI 02904, USA.

Editors

This journal is designed particularly for long research papers, normally at least 80 pages in length, and groups of cognate papers in pure and applied mathematics. Papers intended for publication in the *Memoirs* should be addressed to one of the following editors. In principle the Memoirs welcomes electronic submissions, and some of the editors, those whose names appear below with an asterisk (*), have indicated that they prefer them. However, editors reserve the right to request hard copies after papers have been submitted electronically. Authors are advised to make preliminary email inquiries to editors about whether they are likely to be able to handle submissions in a particular electronic form.

*Algebra to ALEXANDER KLESHCHEV, Department of Mathematics, University of Oregon, Eugene, OR 97403-1222; email: `ams@noether.uoregon.edu`

*Algebra and its application to MINA TEICHER, Emmy Noether Research Institute for Mathematics, Bar-Ilan University, Ramat-Gan 52900, Israel; email: `teicher@macs.biu.ac.il`

Algebraic geometry to DAN ABRAMOVICH, Department of Mathematics, Brown University, Box 1917, Providence, RI 02912; email: `amsedit@math.brown.edu`

*Algebraic number theory to V. KUMAR MURTY, Department of Mathematics, University of Toronto, 100 St. George Street, Toronto, ON M5S 1A1, Canada; email: `murty@math.toronto.edu`

*Algebraic topology to ALEJANDRO ADEM, Department of Mathematics, University of British Columbia, Room 121, 1984 Mathematics Road, Vancouver, British Columbia, Canada V6T 1Z2; email: `adem@math.ubc.ca`

Combinatorics and Lie theory to SERGEY FOMIN, Department of Mathematics, University of Michigan, Ann Arbor, Michigan 48109-1109; email: `fomin@umich.edu`

Complex analysis and harmonic analysis to ALEXANDER NAGEL, Department of Mathematics, University of Wisconsin, 480 Lincoln Drive, Madison, WI 53706-1313; email: `nagel@math.wisc.edu`

*Differential geometry and global analysis to LISA C. JEFFREY, Department of Mathematics, University of Toronto, 100 St. George St., Toronto, ON Canada M5S 3G3; email: `jeffrey@math.toronto.edu`

Dynamical systems and ergodic theory to AMIE WILKINSON, Department of Mathematics, Northwestern University, 2033 Sheridan Road, Evanston, IL 60208-2730; email: `wilkinso@math.northwestern.edu`

*Functional analysis and operator algebras to MARIUS DADARLAT, Department of Mathematics, Purdue University, 150 N. University St., West Lafayette, IN 47907-2067; email: `mdd@math.purdue.edu`

*Geometric analysis to TOBIAS COLDING, Courant Institute, New York University, 251 Mercer St., New York, NY 10012; email: `traneditor@cims.nyu.edu`

*Geometric analysis to MLADEN BESTVINA, Department of Mathematics, University of Utah, 155 South 1400 East, JWB 233, Salt Lake City, Utah 84112-0090; email: `bestvina@math.utah.edu`

Harmonic analysis, representation theory, and Lie theory to ROBERT J. STANTON, Department of Mathematics, The Ohio State University, 231 West 18th Avenue, Columbus, OH 43210-1174; email: `stanton@math.ohio-state.edu`

*Logic to STEFFEN LEMPP, Department of Mathematics, University of Wisconsin, 480 Lincoln Drive, Madison, Wisconsin 53706-1388; email: `lempp@math.wisc.edu`

*Ordinary differential equations, and applied mathematics to PETER W. BATES, Department of Mathematics, Michigan State University, East Lansing, MI 48824-1027; email: `bates@math.msu.edu`

*Partial differential equations to GUSTAVO PONCE, Department of Mathematics, South Hall, Room 6607, University of California, Santa Barbara, CA 93106; email: `ponce@math.ucsb.edu`

*Probability and statistics to KRZYSZTOF BURDZY, Department of Mathematics, University of Washington, Box 354350, Seattle, Washington 98195-4350; email: `burdzy@math.washington.edu`

*Real analysis and partial differential equations to DANIEL TATARU, Department of Mathematics, University of California, Berkeley, Berkeley, CA 94720; email: `tataru@math.berkeley.edu`

All other communications to the editors should be addressed to the Managing Editor, ROBERT GURALNICK, Department of Mathematics, University of Southern California, Los Angeles, CA 90089-1113; email: `guralnic@math.usc.edu`.

Titles in This Series

TITLES IN THIS SERIES

For a complete list of titles in this series, visit the
AMS Bookstore at **www.ams.org/bookstore/**.